Darrell Ason

BASIC PHYSICS FOR BEGINNERS

A Comprehensive Study Guide and Activity Book for the Self-Taught Scientist

© Copyright 2021 by Darell Ason
- All rights reserved -

This content is provided with the sole purpose of providing relevant information on a specific topic for which every reasonable effort has been made to ensure that it is both accurate and reasonable. Nevertheless, by purchasing this content you consent to the fact that the author, as well as the publisher, are in no way experts on the topics contained herein, regardless of any claims as such that may be made within.

As such, any suggestions or recommendations that are made within are done so purely for entertainment value. It is recommended that you always consult a professional prior to undertaking any of the advice or techniques discussed within.

This is a legally binding declaration that is considered both valid and fair by both the Committee of Publishers Association and the American Bar Association and should be considered as legally binding within the United States.

The reproduction, transmission, and duplication of any of the content found herein, including any specific or extended information will be done as an illegal act regardless of the end form the information ultimately takes. This includes copied versions of the work both physical, digital, and audio unless the express consent of the Publisher is provided beforehand. Any additional rights reserved.

Furthermore, the information that can be found within the pages described forthwith shall be considered both accurate and truthful when it comes to the recounting of facts. As such, any use, correct or incorrect, of the provided information will render the Publisher free of responsibility as to the actions taken outside of their direct purview. Regardless, there are zero scenarios where the original author or the Publisher can be deemed liable in any fashion for any damages or hardships that may result from any of the information discussed herein.

Additionally, the information in the following pages is intended only for informational purposes and should thus be thought of as universal. As befitting its nature, it is presented without assurance regarding its prolonged validity or interim quality. Trademarks that are mentioned are done without written consent and can in no way be considered an endorsement from the trademark holder.

TABLE OF CONTENTS

INTRODUCTION	6
CHAPTER 1: What Does it Do?	9
Thermodynamics	12
CHAPTER 2: Movers and Shakers	17
CHAPTER 3: Going Nowhere Fast	32
CHAPTER 4: Up, Down, and All-Around	42
CHAPTER 5: Electromagnetism, Waves, and Light	65
CHAPTER 6: Colors and Light	76
CHAPTER 7: Physics in the Real World	87

INTRODUCTION

Hello! Welcome to physics! I know preconceived notions about the subject probably have you shaking in your proverbial boots right now, and that's okay! Not only is physics a weird word, but it's also an immense and highly convoluted science which deals with just about everything (yes, everything) in the entire world!

If that doesn't have you shaking in the fetal position under your desk, wait; I'm not done yet. Of course, no one can say they know everything about the subject of physics because the possibilities and incredulities are about as vast as our known universe— and probably more extensive than that. But, as I said, physics is literally the study of everything.

Specifically, though, physics in its own right is **the systematic study of the universe and everything it contains**. How things work, how things move, and what things are made of— all these questions can be explained through the application of physics and its specialized mathematical formulas.

It is the basis for a lot of other types of sciences, including biology, chemistry, geology, and even astronomy. These subjects take a lot of their laws and logic from the core concept of physics. They expand upon them, too, in fields such as astronautics and aeronautics (space travel and air travel, respectively). Physics and all of its babies can even be found in the application of computer science and the structure of everything from that built-in solitaire game on your dad's old desktop computer to the furthest reaches and nuances of the internet.

A journey into physics starts with very rudimentary questions (see above— how, why, and what?) and transitions into observations of cause and effect. So if you're curious about anything at all in this world, you can find your answers through physics.

First of all, the word "physics" takes its origin from physis, which in Greek signifies nature. This phenomenon stretches all the way back to the time of Aristotle. He gathered together eight volumes detailing the laws of motion and measurement in the natural world. But he was not the first to apply the ideas of physics to the natural world; there is evidence of the study of the stars and planets as far back as the Meso-American and Babylonian civilizations (250 to 900 CE and 2000 to 1600 BC, respectively).

CE is just a more modern way of saying "after the year 1", so the higher the number, the more recent the period. BC, in contrast, means "Before Christ," or before the year Jesus Christ is meant to have been born.
So, in this case, the higher the number, the further back or older a period is. It's a little confusing, I know, but if it helps, you can also think of CE as "common" or "current era"; the higher the number, the closer it is to our modern-day. Meanwhile, BC is a little easier because, as stated previously, the further back you get from the date of Jesus' birth, the greater the number. It's just the same as marking a year like 1912 and being able to tell how many years ago it was, based on the difference in simple math. For example, 2021 minus 1912 equals 109, so you know 1912 was 109 years ago.

Plato and Aristotle, two Greek philosophers, were some of the first scientists to analyze the movement of objects, but they didn't perform experiments or test any of their theories. However, after his tutelage under Plato, Aristotle went on to expand upon the known aspects of motion and force, using his understanding of the five basic elements:

air, water, fire, dirt or "Earth", and the ether, from which the cosmos are composed.

In his belief that all the elements moved in order to find one another and combine, Aristotle hypothesized that if it weren't for some manner of force, nothing at all would ever achieve movement. Thus some sort of force has to exist for anything to move in any direction, and this was his basis for the belief that there was a direct relationship between an object which moves and that which causes it to move.

After him, there was Archimedes, the Italian mathematician who calculated the mass of an item by observing its movement in water. Later we have Nicolaus Copernicus, the Polish astronomer who blew everyone's minds in the fourteenth century by daring to suggest that the Earth, planets, and stars all revolved in a spherical motion around the Sun. This theory is called heliocentrism, and though Copernicus faced prosecution for it, he died before he could face any kind of punishment.

Then Galileo, whom we'll get into more detail about in a later chapter, came along and confirmed the theory, for which he faced life imprisonment by order of the Roman Catholic Church. Before that, though, like Archimedes, he also dropped things to see what would happen, but he did so from the tops of buildings. And in the very same year Galileo died, Sir Isaac Newton was born. Again, we'll talk more about him in the next chapter, but his name should sound familiar if you have ever heard of his three laws of motion or his theories on gravity springing from an apple that had fallen from a tree.

Okay, no more history and no more math (for now).

CHAPTER 1
WHAT DOES IT DO

Centuries after Aristotle made physics a thing (and after his descendent buddies Archimedes, Galileo, and Copernicus got their mitts on his ideas), modern scientists are still applying a lot of the same rules and experiments today.

One of the most uncomplicated visuals I can give you of physics at work is a lever. Of course, any lever will do if it performs a static upwards and downwards motion, but it might be clearest for the sake of familiarity to picture a teeter-totter. You know, those things on playgrounds that look like a long plank of wood with handles on each end? They sit on some base bolted to the center of the board, and two children sit on either end, grasping the handlebars.

Before anyone moves, both children are on an even keel, perhaps slightly squatting but neither higher nor lower than the other.
When one of the kids pushes off from the ground, they are thrown upwards by the movement of the board. The other kid gets his butt in the sand, with his seat now on the ground and the other guy sitting high up in the air. The action continues, of course, with Sandy Tush doing the same thing the High Guy did, pushing off from the ground and sending himself into the air and the other kid down into the dirt.

In the world of physics, the board the kids are sitting on is the **lever**. Whatever sort of base the board bolted to is called the **fulcrum**.

The **opposing force** is whoever pushes off the ground, setting the lever in motion.

Keeping to the playground theme, consider the humble slide. For this explanation, picture one of those all-metal slides that never coddled you, like the cute plastic spiral slides kids get these days. Instead, the metal slides gave you the chance to get first-degree burns and a face full of woodchips all at once! Wowee!

Anyway. A metal playground slide consisted of a steep, narrow rung of steps and an even steeper, smooth, straight chute. Both slides and steps are examples of an **inclined plane**. The stairs take the brunt of the **load** (the kid) by distributing the kid's weight across its whole surface while at the same time lifting the kid the incline. The climb is also made easier on the kid because they can use their feet separately and in sequence to make it to the top, rather than using all their weight and strength at once to move, like on a ramp.

The slide acts as an opposing force to gravity, in that the friction between it and a jean- or corduroy-clad bottom slows the journey downwards just enough that the kid doesn't go flying into the next county.
Consider the following: Do you ever make use of rechargeable batteries? Do you understand how they work? Did you have any idea they had anything to do with physics?!

Batteries work by converting chemical energy into electrical energy, which is accomplished through the use of their two ends, which are made of two different types of metal, and the conductive liquid, which is primarily composed of electrolytes. In addition, other acids and alkalis are included in the electrolytes, composed of dissolved salt in ionized water and other acids and alkalis.

You might be familiar with two ordinary sorts of batteries: primary (single-use) batteries and secondary (rechargeable) batteries.

Primary batteries can only be used once, while a secondary battery can be used more than once, such as one that can be recharged.

The primary battery functions because the electrodes inside—one known as the anode and the other as the Cathode -negative and positive conductors, respectively— exchange electrons with one another and the surrounding environment. The Anode, also known as the reductor, is responsible for the release of electrons during the powering process.

As the process continues, the Anode becomes oxidized (in this case, used up). The Cathode, which is the positive end of the battery, accepts those electrons and has its power dramatically reduced due to the oxidizing process. As a result, primary batteries can only be used once. The process carried out by the Anode and Cathode in rechargeable or secondary batteries is the same as in primary batteries, with the exception that it is carried out in reverse.

What about those electrons which were transferred from the Anode to the Cathode? In this case, the Cathode returns them to their original state. It's like a miniature ping-pong match between chemicals and electricity!

Regarding batteries and things that require power, headphones are right up there as far as modern 'necessities.'

"What the hell is this nonsense," you exclaim, "there's no way headphones have anything to do with physics!"

Hold on to your seat, and let me bust right through your suspension of disbelief. Sound waves are generated by the speakers in your headphones, which use both electricity and tiny rotating magnets to do so. In this case, the sound waves are emitted from the speaker and bounce directly off your eardrums. Then there's this: your strange little brain takes all of that in and decides it's music. So every sound you hear, from dogs barking to the sound of a gun being fired to your significant other telling you that they love you, is made up of sound waves. When sound waves reach your ears, they bounce off the objects in the environment and travel through the open air to your ears, where your brain processes them to determine what the sound was, where it came from, and how loud it was, among other things.

Within the realm of physics, those sound waves are vibrations, and they can be transmitted through various mediums, including air and jelly. Of course, it will be much easier to hear through the air, but if you make your voice loud enough, the Jell-O will not be able to prevent you from hearing it. Ta-dah! Physics!

Thermodynamics

And finally, for the grand finale, we have your kitchen stove to discuss. Heat causes your food to cook, and the increase and decrease of that heat are made possible by a concept known as thermodynamics, or the study of the change in temperature in any object or material and how these materials interact with one another after being exposed to these temperature changes. When energy flows into and out of our earth system, thermodynamics regulates the flow and conservation of all types of energy. It also saves and distributes some of the energy that is saved and distributed.

In fact, the primary principle of thermodynamics says energy is neither lost nor gained in any type of process, whether chemical or physical in nature; rather, it is conserved by the source of energy. Take cooking as an example: when you cook your food, the heat source is provided by your stove, which can be either gas or electric-powered. If you don't pay attention, the heat will soon spread to the pot, and you could burn yourself on any metal parts that aren't properly insulated. The heat gets transferred from the pot to the food, which is then ready to be eaten. What's particularly interesting about this is that, even though the heat "migrated" from the stove to the pot to the food, neither the stove nor the pot lost any of their energy or heat. They're all on the same level!

And, of course, over time, even with the stove turned off, the burner will become cold, the pot will become cold, and your food will become cold as well. One after another, without a single one being greater or less than the others throughout the entire process.

Thermodynamics also has a lot to do with how people exist from day to day. Or, more specifically, how we have the energy to live.

Humans take in what's called **free energy** in a couple of different ways. One is by eating! You might know that the food we eat gives us energy through proteins and sugars and such, but did you know that the bodily process required to use that food (digestion) and the process which is responsible for converting the food into energy (metabolism) are also reliant on the very same food, in order to work? So really, we even need food to fuel the process which breaks down our food!

And of course, on the external side of things, we humans also know how to burn through a lot of natural resources for energy, such as the use of fossil fuels, nuclear energy, and specific renewable sources of energy such as solar power or hydroelectricity. As we use food to fuel our body's movements, we use these power sources in order to build homes, travel the world, run entire factories, make our own heat, and keep the world lit up even on the darkest of nights.

And while it may not come as a surprise to you, a lot of these resources are finite and cannot be relied upon forever to do what we need them to do. The second law of thermodynamics states the ability for one machine or system to consume energy and continue to operate depends on the availability of the source of its energy to remain constant. Without food to fuel our bodies, we are unable to continue living. Solar panels cannot retain power if there's no sun. If there's no water source available, the mill which takes its power from the movement of a water wheel cannot operate. Things simply can't get done without the energy source we need to do them.

A common topic of many an environmentalist's chants, hippies' songs, and in fact, nature documentaries warns us about the Earth's finite resources, which we continue to use as if they were infinite. Once the source of power and fuel is used up, you get something called **entropy**. This is the use and loss of an energy source without the means to replace it, like burning a campfire and expecting to be able to cook your food the next day on the ashes left behind. You can only burn wood once. All your energy comes from the flame, and when it's burned out, the energy source is gone. Likewise, once all the oil reserves in the world have dried up, it's not like the Earth is going to replace it.

On a lighter note....

The subject of **relativity** is one of those big, scary topics that can scare the more casual dabbler of science to run screaming for the hills.
If you are familiar with Albert Einstein or have done any research on

him, you may know a little about the concept of relativity. If this is the case, you are not alone: even people who have studied it their entire adult lives have only a vague understanding of what is going on. So don't lose heart!

Relativity, to put it as simply as I possibly can, is the dependence of specific events on both the motion of the one who observes them and the movement of the object being observed. If you're still scratching your head, consider the GPS in your car or the one on your mobile phone. The satellites in space that pinpoint your location and the location of wherever you're going do so by moving at a specific speed and sending signals to grounding stations on Earth, which in turn pinpoint your location and the location of wherever you're going. Because recorded time on satellites in space moves microseconds faster than recorded time on Earth, the clocks on board the satellites are set to run the same amount of microseconds slower than the clocks on Earth in order to make them match.

This clarifies my earlier definition of relativity: when it comes to precision, the Global Positioning System (GPS) depends on both the motion and accuracy of the satellites and how fast the Earth rotates.
It is possible for the GPS to be inaccurate if the satellite is not moving as quickly as it should or is moving too quickly. This is because the time measured on Earth will not match up with the time measured by the satellite. In much the same way, if your watch does not show the same time as your friend's watch, one of you will end up arriving late to lunch as a result.

In this chapter, I've used a lot of buzzwords that are printed in bold and appear to be extremely important. Never fear; any of them who aren't fully explained in this chapter will have their chance to shine in one of the subsequent chapters.

All of this is to say that examples of physics in action can be found all around you at all times, and you may not even be aware of it.

For instance, it allows electricity to be produced, gives gold its gold color, improves the accuracy of your phone's map app, and regulates the corrosion of metals. It is responsible for the way these and many other aspects of human life function, and it does so without the majority of us even realizing it is happening.

Isn't that great?

CHAPTER 2
MOVERS AND SHAKERS

Speaking of GPS, did you know that within the science of physics, any body's current location is what's known as its **position**? So, for example, your desk chair might be a foot or two away from your desk, or you might be sitting about fifteen feet away from your bedroom door. There's even a name for the object and your distance from it: the **reference location**. So, where you start from is the reference for how far the destination or position lies away from it.

And then, of course, whatever space there is between you and the reference object is called the **distance**. So when describing the distance between one object and its destination, *plus* the direction in which the object moves, we're talking about **displacement**.

An example of that might be your car, traveling, say fifteen miles to the east from your house. That gives us the object (the car) and the displacement (fifteen miles east). Remember, distance does not include direction, and displacement does; they're two different values. In other words:

Distance = *fifteen miles* **Displacement**: *fifteen miles east*

Before we go into the complexities of movement and the equations and big words which come with it, let's go back to square one.

We'll start with the concept of **motion**. You could also call it 'mechanics,' and it'll mean much the same thing— little rubber bouncy balls to semi-trucks to the great ship Titanic, all of these things require some sort of force behind or beneath them to allow them to move.

You know, if you put your foot on the gas pedal in your car, the car is going to move forwards, and you've probably seen what happens when the white ball in a game of pool hits all the other balls on the table, causing them to scatter.

Your foot, or the cue stick, represents another influencing phenomenon known as **force**. Anything that changes the speed or position of an object is considered force, whether that's by the action of pushing (like starting a boulder rolling down a hill) or pulling (dragging your dog away from that fascinating pile of roadkill it's found in the road).

If you put a bowling ball on a sturdy oak dining room table, the table will not give under the ball's weight because it is capable of exerting enough force to hold up the bowling ball. However, if you hold the bowling ball in one hand, your hand will be pushed downward because your hand is not as strong as the table and cannot produce the amount of force it takes to hold it up completely. This, of course, depends on the weight of the bowling ball and how strong your wrists are. But with pretty much any weight of a bowling ball, if you put it onto a sheet of paper towel and try to lift the ball with the towel, the ball will easily break through. It is not capable of exerting force strong enough to hold up the ball.

These are called **contact forces**, wherein one object exerts a certain amount of its force to either hold or moves another object. The weight of each will determine if the movement is possible or not or if the first object will completely fail to hold or move the second and will break.

This is the road under your car, a kayak being pushed downstream, or a windsock on a windy day.

If you drop any kind of ball from any height, it gains speed as it falls until it makes contact with the floor. But what if, instead, the ball rolls off of a table and falls onto the floor of its own volition? Spoiler alert: nothing really is changed.

In both instances, gravity works downward force upon the ball from a vertical direction. However, in neither case is the ball pulled to either side by any force acting horizontally, and so if the only changed value is from where the ball falls, it will still come in contact with the floor in the same instant, either dropped from your hand or rolled off of a table.

Now, to mention our old friend Galileo again, he had his own theory of relativity, which was completely independent of Einstein's theories of either special or general relativity. Galileo's theory helps explain the variables of the above hypothetical in a little more detail and gives us a good example of what would happen if there were some sort of horizontal force at work on the ball.

Again, as far as speed or actual movement, his experiment doesn't vary much from just dropping the ball from shoulder height. But it does prove something interesting with regards to the laws of motion.
If a sailor stands in the crow's nest of a moving sailboat and drops the ball, the ball drops straight down according to that sailor's line of sight.

If someone were watching from shore, however, to them, it would look like the ball was thrown or dropped in an arching motion or that it remained up in the air longer than it did. This is because while the sailor is on the ship and therefore moving at the same speed, the observer is standing still on land. And though the path of the ball would look different to the respective onlookers, they would both still observe the ball hitting the surface at the same moment.

This is because, according to Galileo, the laws of physics are autonomous of any type of dependent motion. So basically, it doesn't matter if you're on a boat or on land or flying in a helicopter over the Grand Canyon; physics will work the same way, and that ball is going to continue to accelerate and fall straight downwards until a floor, or anything else stops it.

Now. If you throw a baseball like you're throwing a pitch, that ball is going to take an upwards and outwards path, not a straight across one. And the distance traveled before it's stopped by either a glove, a fence, or the ground, directly depends on its speed and what sort of angle it was launched in.

In the case of a baseball game, let's say the pitcher threw the ball at around 98 miles an hour. That gives us 132 feet per second. Of course, the angle depends entirely on the pitcher's skill; say he's on the top of his game that day and throws the ball at an angle of about 25 degrees. At that angle, the ball is destined to go right to the batter's center of vision, and once hit, will maintain a constant speed even as it rises and falls, though the angle of its path would have changed at impact.

That's it for the sports metaphors— let's try something a little bit more fun. Picture a carousel in motion. All of those horses and benches and various other animals can only move in a circle, right? And once the carousel gets up to speed, it stays at that same speed for the duration of the ride. Same speed, same direction, only ever in a perfect circle; coincidentally, this phenomenon is known as **circular motion**!

The force, in this case, is the engine or whatever powers the carousel, and it only over pushes the ride in one direction, with no opposing force working against it. And even though it's only moving in a circle, its velocity (speed and direction) changes as soon as it starts to move.

So the force required to keep the whole thing turning is dependent on a number of things: the density of the body, its velocity, and how large or small the circle in which it moves is (aka the radius). Of course, the force has to be stronger the larger the thing is and how fast you want it to go, but interestingly enough, more miniature carousels actually require more force to keep at a constant speed than do the larger ones.

This is because the sharper the turn, or in this case, the tighter the rotation, the smaller the turning circle. This is what's known as **centripetal force**. As a result, the engine requires more force to keep you and the machine moving on a child-sized carousel because there is less distance between you on your horse and the force trying to move you.

The centripetal force equals the classification of force that triggers any movement of something in a curved direction. Whenever it moves, it moves in an orthogonal, or independent, direction with regards to the motion of the body and in the direction of the fixed point of the direct focal point of the bend of the path, i.e., around an engine of the carousel, which is almost always dead center of all the moving horses. Isaac Newton defines the centripetal force as any sort of power by which something is drawn towards or repelled away or otherwise moved by no effort of its own. According to Newtonian mechanics, gravity is responsible for the centripetal force that causes astronomical orbits, like the planets around the Sun. So the carousel's motion around the force that pulls and drives it, the whole platform and everything on it, is directly dependent and tied to the power source in the middle.

If you recall the concept of centripetal force from earlier in chapter two, you will remember that something rotates around a central force as a result of a gravitational or physical pull. It is possible to simulate gravity by spinning a ball on a string in a continuous circle above your head. The string acts as an anchor, keeping the ball anchored to your hand, which is the driving force behind it.

Similarly, the Sun generates a gravitational field that keeps the planets tethered to it, causing the planets to revolve around it in a constant circle. The planets, in contrast to a simple ball, do not all move at the same speed or follow the same path in the same direction.
This is primarily due to the fact that each planet has a mass that is entirely unique to it. Furthermore, the path of each planet is shaped less like a circle and more like an oval the further away from the Sun it is traveling.
To throw another name into our ring of scientists and discoverers: Johannes Kepler, a German philosopher, and mathematician studied the observations of Mars put forth by yet another man, Tycho Brahe. Using Brahe's basis for the shape of the planet's revolution, Kepler went on to confirm— after many failed attempts— that Brahe had been on the right track, after all. His conclusion that the shape of orbit was more an ellipse than a perfectly round circle gave credence to what would later become his first law of orbital motion.

Similar to Newton and his three laws of motion, Kepler devised his own three laws dictating how exactly the planets move around the Sun. Later, of course, it was found that these three laws could be applied towards the orbiting movements of any object around another due to gravity and didn't only pertain to the orbit of planets in the solar system.

As was previously thought, Kepler's First Law of Orbit states that planets orbit the Sun in an elliptical shape and not a circle. Instead, the

Sun sits in the center, with the respective planets rotating on an alternatively widening and narrowing path. As a result, whenever a planet is nearer to the pull of the Sun, it will move more quickly than it would in its outer loop, further away.

Kepler's **Second Law of Orbit** explains this previous point of speed of planets while also interceding that once the planet's orbit is set and balanced by the Sun, it will keep the same rate of acceleration and slowing down according to its place on the loop and that these values will never change.

Kepler's **Third Law of Orbit tells** us that the calculated gap of any sphere from the Sun, cubed (or taken to a power of x^4) directly correlates to the time it takes for that planet to complete its orbit squared (taken to the power of x^2). So, put in plain English, the distance at which a sphere is located away from the Sun sets the time it takes for that sphere to circle the Sun.

Satellite	Type of Orbit	Altitude (km)	The radius of orbit (km)	Period of Time
International Space Station	Equatorial	278 to 460	6,723	91.4 minutes
Hubble Space Telescope	Equatorial	570	6,942	95.9 minutes
Weather Satellite (NOAA)	Polar	860	7,234	102 minutes
GPS Satellite	Equatorial	20,200	25,561	718 minutes (approx. 11 hours and 56 minutes)

Communications	Equatorial	36,000	42,105	1,436 minutes (approx. 23 hours and 56 minutes)
Moon	Ecliptic	364,397 to 406,731	384,748	27.3 days

The third law, according to Kepler, is also indicative of the movement of certain orbiters around Earth. In the way that the planets circle the Sun, these satellites are programmed to move at a set distance away from the Earth, depending on what they are measuring and/or recording.

Objects like the weather satellite have **polar orbits**, meaning that they cover a different area on the Earth's surface every time they make a complete revolution. This is because of the tilt of the Earth and the speed at which it rotates. This is why a weather satellite is able to monitor most of the planet at a time, thus giving up-to-the-minute readings of whatever is happening down below, as far as storm movement and cloud cover.

The Moon follows an **ecliptic orbit**. The Ecliptic itself is a circle representing the Sun's path of movement over one Earth year. So it is so-called because lunar and solar eclipses occur only at an instance where the Moon crosses into this path.

Despite the fact that Sun measures about 400 times bigger than our Moon and about 400 times the distance from us, they appear roughly the same size as someone standing on Earth. This is why solar eclipses are possible. If the Moon were further away or smaller, there's no way it'd take up enough space (ha!) to cover the Sun (in our perspective) completely. Then again, if it were larger and closer, it would block out the Sun on a much more regular basis, which would make eclipses a lot less exciting.

If you remember, I told you that the planets' orbit around the Sun is elliptical, i.e., not a perfect circle and more like an oval. Because of this, we're closest to the Sun in the first few weeks of January and furthest away at the beginning of July. Sounds backward, doesn't it? But the human eye can't register it anyway, so don't feel too bad. Keep in mind, too, that the Earth is also spinning on its axis while rotating around the Sun. This is why different sides of the world have their seasons at different times; Australia has its summer when the continental United States has its winter. But, of course, it's pretty much always hot in Australia, and they have spiders the size of dinner plates, so if you ask me, that's a pretty fair tradeoff.

Anyway. Believe it or not, the Moon appears 14% larger when it's closest to Earth than when it's at its furthest away. So at its closest, it can easily cover the 'solar disk' and create a solar eclipse because of its position between the Sun and us.

Likewise, the tilt of the Moon's orbit sits at a five-degree angle to the route of Earth about the Sun. This, by the way, is also a definition of the **Ecliptic**, mentioned a little bit ago. The Moon will either pass above or below the Sun in its orbit, preventing the phenomenon of an eclipse from happening. But twice a year, the Moon passes just close enough to the Sun's path to cause at least a partial eclipse, which does not last as long as an annular or total eclipse.

Roughly every 18 years, this elliptical cycle repeats, and every year the path of the moon shifts west equals the length of one-third of the Earth's rotation. It will continue to change little by little until a total solar eclipse occurs, and then around and around again, moving closer and closer again until it happens again, in the exact location, another 375 years later.

There are in-between stages, of course, like the Blood Moon, which is when the Earth blocks light from reaching the Moon. This can only happen on the occasion of the Moon, Sun, and Earth being in a straight line with one another, with the Earth in between. However, these are much more common than a total solar eclipse, occurring at least four times within two years.
And it's called the Blood Moon because the shadow of the Earth turns it a dark, rust-red color. Spooky!

Now, coming back to the subject of force after a long space tangent. But you learned something, didn't you? And all of that was still related to force and gravity, just on an extremely grand scale.

Related to the concept of force, of course, are the concepts of **acceleration** and **deceleration**. Think of a car. Better yet, think of a minivan and a sports car. They're the same age, both in prime condition, but the sports car is going to increase its speed from zero to 60 miles an hour in a much shorter time (around 5 seconds) than the minivan, which might require as much as 8 seconds to reach the same speed. One term I need to introduce to you here is that of **vectors**.
We'll talk more about them and what they entail in the next chapter, but they're essential here because they pertain to acceleration.
Both velocity and acceleration include descriptions of both magnitude and direction (how fast and how far). This makes them both known in the physics world as vectors, which are important values to understand when it comes to these types of equations.

If you have any previous knowledge of physics before opening this book, you might be familiar with a man named Sir Isaac Newton.

He was an English "natural philosopher" (which at the time is what they called someone well-versed in natural and physical sciences). Basically, he would be a great guy to have on your team at trivia night.

Newton's contribution to physics included the ways in which the movement of items in the universe correlates to the object's weight, speed, and force; basically, how, why, and how fast an object is moving. A touchstone of basic science includes his three laws of movement, which include the following:

Newton's Law of Inertia tells us that if zero amount of push is put to an object while that object is resting, the object will continue to rest.
On the other hand, if the object is in motion, it will proceed in its motion at a constant speed and in a constant route to the one before.

Newton's Principle of Constant Velocity states that the movement and speed of any item are proportional to the intensity of any force acting upon it (or vice versa). The size and direction of the force have an impact on the velocity and direction of the object, as well as the direction in which it is moving.

Newton's Law of Action and Reaction states that there will always be an equal and opposite reaction to every action.

We'll start by exploring the first law. Inertia is an object's resistance to acceleration or movement. The expectation is that the object will remain unmoving or transfer at the same velocity in a single direction until it is interfered with by some type of force.

You can find an example of this almost anywhere in your daily life. One instance might be a bus ride. If you're one of the people standing up and holding on to one of the poles, you're at the mercy of the motion of the bus much more than someone seated. If the bus makes a sudden, unexpected stop, odds are your body will go flying forward, and it will continue to fly forward until you come in contact with either the dashboard or the floor or another rider.

The bus stops, but your body, which up until this point had been moving at a similar speed as the bus, does not have the benefit of brakes or the resistance of the road on the bus tires. You will continue to fly forward until something stops you.

However, if you happen to be free-floating up in space, the rules for you are a little bit different. First of all, God's speed and good luck. Because even the planets and the moons obey Newton's law of action, not accounting for the gravitational pull of our Sun, they'd all be moving in straight lines, on their own little conga line out into the eternities of space. But because the Sun pulls them towards itself, they will instead rotate in wide arching circles forever, unless something somehow gets in their way. Same with an astronaut, if they come untethered from their umbilical lifeline and no longer are held at bay by their ship. Like the Duracell bunny, they just keep going, and going, and going...

Newton's second law involves the concept of velocity. Let's say you're shopping at a hardware store. You plan to buy a lot, maybe to revamp the flower garden in front of your house. So you and your companion take two carts. They load theirs up with heavy bags of mulch, maybe some nice edging, a new shovel, maybe a big bag of birdseed for good measure.

Meanwhile, your cart has only a couple of plastic watering cans, some cute pinwheels to catch the wind, and maybe a couple of garden gnomes. If you're in a hurry, the two of you might speed up and make a beeline for the registers. Once you do, two things are going to happen here.

Your lighter load is going to allow you to speed up more quickly than the person with all the heavy stuff. Once you reach the registers, it'll also be easier for you to stop. If, by chance, your companion manages to push their heavy load and get up to speed, they will not be able to stop on the dime.

Newton's law of constant velocity is at work here in a couple of ways. One, you both exert force upon your carts and powerwalk towards the front of the store. The exact amount of force is not needed to push your cart as is necessary to push the heavy cart. The relation between the mass of the object (the cart) and the amount of force applied against it (you versus your friend) makes a difference in the acceleration produced by your pushing. Adversely, the increased weight of the other cart makes slowing down and stopping harder on a flat surface because that mass is pushing itself forward.

Guess what? This is still true in space! Even without gravity, enacting the same amount of force upon lighter or heavier objects will result in different speeds of those objects because weight still applies. Think of it this way: it'd probably be way easier to play ping-pong in space than it would be to play basketball.

The Third Law of Motion says: with any power acting towards an item, another identical power acts in the opposing path in order to keep the object moving. Therefore, because one thing happens, another thing happens. For example, if an animal sits on a stationary object, and her weight applies any amount of energy on the chair, the plane on which she sits also applies an upward surge on the animal, which holds her upright. Similarly, and keeping with our space theme, the power applied to a planet by our Sun is also "felt" by it; but due to the fact that the Sun is so much larger than any of the planets, there's little effect on the Sun's motion, despite the fact there will be a greater impact on the planet.

Likewise, if you hit a nail with a hammer, the nail will go into the piece of wood. Both the nail and the hammer are 'aware of' or affected by the blow, but the nail is most drastically altered and feels more force than the hammer, which is larger and heavier. Likewise, if you hit a baseball with a bat, it's going to fly through the air. If you've ever actually swung at a ball and connected with it, you know there is a funny feeling of vibration in your arms when ball and bat meet at that high speed. You might've already guessed, though, that the ball feels a lot more of the brunt of that impact than either you or the bat. Same thing! Simple, right?

For the last bit of this chapter, I'm going to briefly introduce you to the concept of gravity. We'll get a more in-depth look at the functions and forces of gravity in chapter four, but since we're talking about Isaac Newton here, I figured we'd let him have one last moment of glory.

As you might already know, **gravity is the action of a force between two objects that pulls them towards one another**. In fact, **Newton's Universal Law of Gravitation** includes gravitational influence between entities in space, such as the Earth and its Moon or Earth and the Sun. But most famously, Newton came across the phenomenon of gravity when he observed an apple falling off of a tree. It did not, as the legend says, actually hit him in the head, but the fact that the apple fell straight down rather than in any other direction gave Newton pause.

He went on to publish his ideas on gravity in 1687 in his larger work entitled the "Principia," which also included his three laws of motion.

His initial statement put forward the idea that everything in the universe is attracted to everything else in the universe, and the force's strength directly depends on the weight of the respective objects and the distance between the two.

Albert Einstein had a lot to say about gravity as well, but we'll save his thoughts for a later chapter so that we can give him his due. From here, we can move on from the sensation of movement to the relativities of scale and distance, which in their simplest forms are known as vectors.

CHAPTER 3
GOING NOWHERE FAST

If you've ever given directions to someone or helped someone hang a painting on the wall, or swerved to avoid hitting an animal in the road, you've used vectors. That's because vectors are things like measurements or magnitude (distance) and directions. For example, left, right, north, south, five miles, ten kilometers— etcetera. And in the world of physics, when you're talking about where an object is before it is moved, you're talking about its position.

We touched on this briefly in chapter two and how the difference between distance and displacement is the specificity of the description. Distance would be saying something is fourteen miles away from you. Displacement is saying something is fourteen miles to the east of you. See the difference?

When you take a value like this, such as saying your house is fourteen miles to the east of the nearest McDonald's, you're describing a vector.

As they pertain to physics, Vectors are a little more complex, exploring concepts such as acceleration, velocity, and force. But they are a pretty fundamental part of this particular branch of science, so it's best to get as comfortable with them as you can.

As I've already said, the two main qualities of vectors are direction and magnitude. But you can also have vectors with only one quality, such as speed. These types of vectors are usually referred to as scalars. Take a scalar and add a particular direction; you will get a vector. There are also null, or zero, vectors, which basically denote an object's velocity when it is, in fact, standing still.

Negative vectors are two vectors of equal magnitude or length but which head in different or opposite directions. Something like this:

Co-initial vectors are two vectors with a common origin point. They will still travel in slightly different directions, at different angles, but they relate from the same coordinate first.

Collinear vectors are those with different magnitudes (or lengths) but still acting along the same or parallel path.

Position vectors give the position (go figure) of an object by referencing its point of origin within a coordinate system and are most typically represented by the symbol *r*.

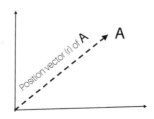

And lastly, a **displacement** vector is the type of vector which tells how far and in what direction an object has changed its position in any given passage of time.

So again, an example of this might be fleeing a bank after you've robbed it, in a van in which you shouldn't be putting too much faith to get you where you need to go. Say it takes you about 22 minutes to get 15 miles away from the scene of your crime and allowing for mid-day traffic, you're driving at about 40 miles per hour. Not ideal for a speedy escape, but hey, at least you're a conscientious driver.

So you're going 40 miles an hour for 22 minutes and make a distance of 15 miles. So all of these things together make up a displacement vector.

To visually represent a vector, the most popular method is to draw an arrow. This is because arrows represent a clear *direction* and *magnitude* (however long the arrow is). So like this:

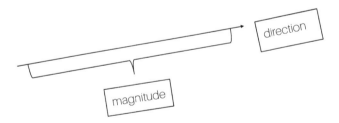

Vectors can represent force, acceleration, velocity, and other units of movement. Usually, within a text, the capital letter **A** is used to represent a vector. Sometimes, an arrow is drawn over top of it to show that the vector has a scalar or speed value and a direction. It's easier than drawing out a diagram like the one above!

Similarly, if two vectors are represented and look to be identical, this can be represented by the simple equation of **A** = **B**. Or :

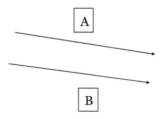

Same direction, same length; thus, the two vectors are equal.

These are, in fact, called **equal vectors**. Now say someone is giving you directions to a specific location, but it's not a matter of driving in a straight line. For example, that bar you and your friends are meeting at for your fifteenth high school reunion might be twenty miles west, then another thirty miles south. But wait, that's two different directions! What do I do?!

Well, in the real world, you drive twenty miles, make a turn, and drive another mile and have a grand old time. Within the realm of physics, however, you sit down with a pencil and paper. It's a simple addition problem. And when you add the values of the two known locations together, you get something called the **resultant vector**, which is just like the sum in regular addition, except that it has a distal value.

So **A** + **B** = **C**, or the first vector (twenty miles west) plus the second vector (thirty miles south) equals the total distance you've come from your starting point.

So you drive the distance of vector A, make a turn, and drive the distance of vector B. To figure C, you draw an arrow or a line from the initial point **A** to point **B**, like so.

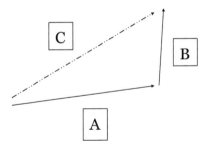

Vector **C** represents the whole, or sum, of your trip from start to finish. But wait, there's more! You can also subtract vectors, like taking the distance traveled (C) minus one of the other values (A) to find the length of the second leg of your trip (B). You can use the same diagram as above to do this, but fill in the already known value of C and A and leave B blank. Without the fancy letters and diagrams, it's just the same as regular subtraction to find value.

C − **A** = **B** is quite literally the same as 100 − 47 = 53. In this case, the value of 53 would be your C, 47 would be A, and the solution of 53 (B) would give you the second shorter distance needed to make the sum equal 100.

Now let's talk graphs! You may have used numerical graphs and coordinates in your high school algebra and geometry classes, and you might be tempted to breathe into a bag at the very idea of revisiting them now, but don't panic too much. The whole point of this book is to make each and every one of these concepts as pain-free and simple as possible, so that you don't see lines and crosses and grids and run straight for the hills (anymore).

See the graph below.

	A	B	C	D	E	F	G
1							
2							
3							
4							
5							
6							

To keep it simple, let's use a measurement of inches for this exercise. So vector **A** is one inch up and five inches over, and vector **B** is one inch to the right plus four inches up. If you want to add them together to get vector **C**, combine the horizontal coordinates and the vertical coordinates together.

> A's coordinates look like this: (5,1)
> B's coordinates look like this: (1,4)

So to get C, it's as simple as adding A+B or (5,1)+(1,4) to get C, or (6,5).

So you see, once you break it down like this, it really is identical to simple addition or subtraction. You just have to seek out the numbers and pinpoint the equation needed to get the correct answer.

Now, take a minute. Relax your muscles. Take a couple of deep breaths. I will hold your hand and use small words, and we will get through this together.

Good? Okay. We're now going to learn how to *multiply* vectors.

Say you're the competitive and slightly reckless sort, and you're driving along on the freeway at a comfortable 60 miles an hour, heading westward. Some moron in a souped-up sports car comes up beside you in the other lane, obviously looking for a challenge. You're not worried about it, but you do have your pride.

Apparently, under the impression that your better angels are watching you like veteran Navy Seals and wouldn't allow anything bad to happen if you make a stupid decision, you decide to push your car to the limit, double your speed, and leave this idiot in the dust. In mathematical terms, this brush with death would look like this:
$$2(0,60) = (0, 120)$$

Now you're positively flying down the freeway, going one hundred and twenty miles an hour, and praying any cop you see is driving a Crown Victoria and not a Dodge. This is a good illustration of the multiplication of a vector (2x speed) by a scalar (60 miles an hour).

Okay, now that we know where you're going, what direction you need to travel in, and how much of a distance there is from point A to point B, let's talk about how quickly you're going to get there. You know about speed. But do you know the difference between average and instantaneous speed?

The definition of **speed** is the rate at which the change occurs and how much distance was gained within that time. So suppose you're on a road trip, and by the end of your day, you've traveled 350 miles in five hours. The straight mathematical solution is that you traveled at 70 miles an hour. But because the average person does not and cannot remain at a constant speed, unchanging, for that long, the 70 miles per hour is considered your **average speed**. It's not actually how fast you traveled for the entire trip.

Instantaneous speed is the threshold of space divided by a point in a moment when the time is reduced gradually until you get 0. Put less intimidatingly, think about how your speedometer on your car tells you how fast you're driving. The speedometer uses the torque or force, put upon a disk made out of aluminum. The disk is pulled in rotation by an extremely strong magnet, which in turn is rotated whichever way the car axle turns.

So unlike a computer or a calculator determining the average speed you would need to maintain in order to make a trip in a certain number of hours, the speedometer and its related mechanisms literally use the movement of the car to determine precisely how fast you're going.

Now, **velocity** is the combination of speed and direction, specifically the dislocation, then divided by how much time it takes in order for the change to occur.

$$\frac{\text{displacement}}{\text{time required for change}} = \text{Velocity}$$

Remember what we talked about before, when we talked about displacement? When it comes to measuring the concepts of displacement and velocity, the two are very similar. They are both a measure of the distance traveled plus the direction in which the object moved, like a car driving twenty miles to the west.

In the process of learning about physics, it is common to make the mistake of assuming that displacement and distance are of equal value or that they are interchangeable terms. Unfortunately, this is not the case.

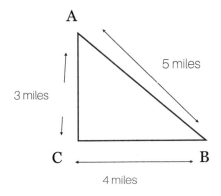

Take a look at this diagram. When walking from A to B via C, or by walking south and then southeast, your total distance traveled would be 3 miles plus 4 miles, for a total of 7 miles. However, if you omit the three miles between A and C and instead travel directly southeast from A, you will still arrive at B and will have saved two miles on your journey!

In short (another pun!), distance is the total length of a journey traveled from home, hitting the rest stop in the middle, then walking the rest of the way to your destination. A to C to B.

Displacement is the shortest possible journey between your starting point and your final destination, i.e., A to B. Knowing that distance must always be positive. In contrast, displacement can be positive or negative, or even zero, dependent on the trip and the direction, which is yet another way to differentiate between distance and displacement. So, once again, it's time for a diagram!

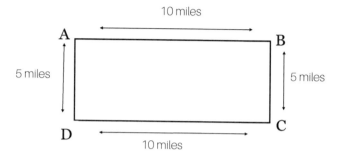

You can calculate your total distance by starting your car at A and driving it clockwise around the perimeter. For example, if you start at A and drive through B, C, and D in a clockwise direction until you return to A, your total distance will equal the perimeter of the rectangle (or cornfield?): 10 + 5 + 10 + 5 = 30 miles.

In this case, your displacement is zero because you have returned to the point from where you began. But if you drive from point D to point A for that distance of 5 miles, you will have been displaced a total of -5 miles further than you started. And it's negative because, in this case, both the direction and the quantity make a significant difference. So in an equation like this, moving upward (from C to B) results in negative displacement, while moving downward (from B to C) results in positive displacement.

Suppose you decide to go rogue in your little car and drive an additional five miles upwards from A; your total distance, in this case, is 35 miles because it includes the total distance from A to B, C, D, back to A, and then an additional five miles in the path of negative displacement beyond A.

CHAPTER 4
UP, DOWN, AND ALL-AROUND

Okay! Finally! When I teased you about gravity a couple of chapters ago, you might have assumed that hearing a description of how Isaac Newton watched an apple fall out of a tree would be the most exciting thing we'd get to talk about. But you'd be wrong!

Sir Newton did, on the other hand, provide us with an excellent jumping-off point. (look, it's a pun!). His first significant contribution to the concept of gravity is the law of universal gravitation, which he wrote about in 1687. Essentially, everything in our universe is attracted to everything else in the universe, and this is known as the attractive force. But, of course, you are already aware of this because your retention is excellent, and you have been taking notes, correct?

This theory becomes significantly more complicated once you leave the planet Earth. Still, if we continue to follow Newton's law, which also says that the power of gravity between two items is proportionate to the size and weight of the objects, we should be fine.

Consider the Earth and its Moon as an example. Due to the fact that the Earth is approximately four times larger than the Moon, our home planet's pull upon the Moon is approximately 81 measures more powerful than the Moon's gravitational pull on the Earth.

So now we know the Moon revolves around the Earth rather than the other way round. Similar to the way that the surface of the Moon bulges towards the Earth, large bodies of water also bulge away from the Earth and towards the Moon. Different bodies of water around the world rise and recede at different times of the year, depending on where the Moon is in the sky at a specific time. That's why we have waves!

In a similar vein, the gravitational pull of the Sun has a varying effect on the different planets, depending on the density and composition of each planet. Furthermore, although the fact that the Earth is roughly four times as large as the Moon, the Sun is 109 times larger than the Earth! Thus, the reason why all planets don't orbit in the identical direction and why they are all unevenly spaced from one another is due to the fact that their densities differ from each other.

A planet's mass is proportional to how strongly it attracts the Sun's gravitational pull, which results in a closer orbital path for the planet. The planets designated as inner planets are those nearest to the Sun. This includes Mercury, Venus, Earth, and Mars. They are relatively small in size and composed primarily of rock and various types of iron and nickel. Because they are smaller and denser than the rest of the planet, they are more sensitive to the Sun's pull.

The outer planets, which include Jupiter, Saturn, Uranus, and Neptune, are so named because they are the planets that are the furthest away from the Sun. They are also composed of rock and metal at their cores, but their outer bodies are composed of hydrogen and helium, two extremely light gases that make up the atmosphere. So, despite the fact that these planets are the largest in the solar system, they are also the least dense, allowing the Sun to exert a less gravitational pull on them. As a result, they're much, much further away.

Looking back at the concept of centripetal force from way back in chapter two, you'll remember that something rotates around a central force because of a gravitational or physical pull.

If you spin a ball on a string in a continuous circle above your head, the string acts like gravity, keeping the ball tethered to your hand, which is the force that drives it. So likewise, the Sun produces a gravitational field that keeps the planets tethered to it, keeping the planets moving in a constant circle around it.

Unlike a simple ball, however, the planets do not all move at the same speed, in the exact same path. This is due primarily to the mass of each planet being unique to it. And the path of each planet is shaped less like a circle and more like an oval the further away from the Sun.

We can use Newton's principle of motion for the theory of **gravitational fields**. The gravitational **push** on something depends on the density of two items split by the square of the expanse between them. So, for example, the most recently known mass of the Earth is 5.9736×10^{24} kg or 1.31668×10^{25} pounds. If those numbers scare you, that's about 13 billion trillion tons. She's a big gal.

Now, the Moon has a mass of roughly 7.33×10^{25} g = 7.33×10^{22} kg, or 8.1×10^{19} tons, or thirty quintillion nine hundred quadrillions. Don't try to make those numbers make sense. It's okay. Remember that the Moon's mass is less than that of the Earth, but if you're curious, the answer to the equation that tells you the gravitational power put on the Moon by Earth is $F = [Gm_1m_2/d^2]$.

Hold on a sec; there's a little more math to learn. We started talking about the gravitational field of a particular planet.

How much gravitational force is exerted on the planet varies by the amount of power being applied to the planet. The value of a planet's gravitational field is calculated by dividing the gravitational force by the size of the planet in question. The gravitational field is represented by the symbol g and the universal gravitational constant (or the proportion of non-relative gravity given to all objects in the universe).

I won't try and explain too much about relativity, which is a concept discovered by Albert Einstein. For the sake of explaining gravity fields, however, I'll describe it as simply as I can. Einstein determined space and time to be relative, and all the action must be comparative to a form of reference. Or even more simply put: all the laws of physics are the same everywhere, forever and ever, amen.

So, non-relative gravity is basically the concept of everything in the universe having its own, different gravitational field. Don't bring this up with Albert.

Speaking of, Albert Einstein is the last but certainly not the least in our list of influential minds in the worlds of physics and other sciences. Born in Germany, Einstein established his first ideas about spacetime, EM fields, and statistical physics during his researches at the University of Zurich. One of his earliest theses was on the subject of special relativity (there's that word again, yikes!), exploring new theories and equations pertaining to extreme rapid speed (such as that traveled by rocket ships) and electromagnetism, which is the study of electric currents and their effect on magnetic fields. Perhaps most famously, however, he published his theory of relativity in 1916, at the age of thirty-seven.

I promise I'm almost done talking about relativity. But in this case, it is relevant to what we've been talking about in this chapter.

One of the predictions within this theory is as follows: a star's light doesn't have to move in a straight line but can take on a curve if it moves close enough to a powerful gravitational force, such as the Sun. If this sounds familiar, good! You've been paying attention.

Newton's theory of universal gravitation, which I discussed earlier in this chapter, essentially laid the foundation for Einstein's more detailed discovery, which stated that every object in the universe has an effect on a gravitational pull on every other object in the universe. The fact is that while Newton's theory acknowledged the possibility of light being bent due to the force of a gravitational field, he did not allow for as much magnitude of that force that Einstein did, and when the Sun went into eclipse in 1919, Einstein was able to test and prove the facts of his own hypothesis.

Although the occurrence was recorded by Arthur Eddington, an English astronomer, and fellow physicist, it was not until much later, when it was more famously explained by Einstein, that it became widely known. A particular cluster of stars normally appeared at night, every night, in the same location in the sky and at the same distance from the Sun, with no variation in their appearance or position. In order to prove Einstein's theory correct, Eddington had to be able to demonstrate that the Sun's movement in the sky on the night of the eclipse would push this cluster of stars further away from the Earth's atmosphere. Due to the fact that the Sun's light is too bright for these stars to be visible during the day, they would only be visible during a solar eclipse. But, of course, the gravitational force of the Sun did shift the position of the stars, pushing their visible light away from itself, thus proving Einstein's principle of general relativity that gravity can bend light. Never mind that he had to course-correct from Newton's findings, give or take a few quadrillion kilograms!
By the way, the concept of a gravitational field is a vector quantity.

It all comes back around! Gravity fields are the force acting on an object (a planet) that causes the object to increase speed in the direction of the pull acting on it (the gravity field). This is the reason why the planets that are closest to the Sun move faster in their orbit than the planets that are farther away from the Sun. The term **gravitational acceleration** refers to this concept, which should be well-known among physics students.

Here on Earth, the numbers required to explain gravity's effect on everyday objects such as an apple or an elephant are much smaller than those required in space. And, of course, we can't talk ups and downs or throwing heavy things to see what happens without talking about Galileo Galilei, an Italian physicist and engineer who came before Isaac Newton.

Galileo's initial experiment with weight and mass in regards to gravity was very simple. Starting from the top of the Leaning Tower of Pisa, he dropped various items of different loads and mass (for example, a cannonball and a ball made of wood) from the same height.
He observed that each object took equal time to drop, and they all hit bottom at the same moment as whatever they were dropped with. With this, he was able to establish a rate of acceleration and that the speed at which an object falls has nothing to do with its mass (see the cannonball and the wooden ball).

While it may not matter how much mass an object has when observing the speed with which it falls, the one time it does matter is when you're looking for an object's **terminal velocity**. Terminal velocity is the greatest speed reached by an object as it falls. This is accompanied by a phenomenon called air drag. Also referred to as drag force, air drag is an influence acting in the opposite of the motion of any body going through a solution or the air. The quantity of air drag on an entity also varies by the density of the air (or liquid) through which the object is passing, as well as the size, shape, and initial velocity of the object. The terminal velocity is determined by taking this information, plus the object's mass, into account.

A couple of real-life examples of air drag and terminal velocity at work are parachuting and going down a water slide. If you're parachuting, aka jumping out of a plane and hoping a bit of polyester will stop you from splatting on the ground, you keep your arms and legs tucked into your body and your parachute in your backpack, up until a certain point after the jump. You're freefalling. Then, at the exact right time, you pull the cord which opens the parachute.

As soon as you've done this, you're yanked back upwards in the air for a moment, and then the rate of your fall is dramatically reduced by the open parachute, giving the air an obstacle to push through. Obviously (thankfully), it can't slip through the polyester with the greatest of ease, so you float safely to the ground.

Similarly, when you go down a waterslide, the speed at which you zoom towards the bottom depends on how you sit at the top. Either way, of course, there's the water on the slide providing minimal drag against your body, but it's not going to slow you down that much. If you sit up and throw your hands up in the air, you're causing more resistance, more drag, against the gravitational pull of the downward slope. If, however you decide to lay down with your arms tucked in at your sides, you're going to get to the bottom much more quickly. You've drastically reduced the amount of drag with which gravity and force must contend because there's less of you (in a spatial sense) which needs to be moved.

As far as terminal velocity goes, it ties in directly with weight, which relies on gravity. When you weigh yourself on the scale in your bathroom, the number it displays represents the amount of upward force the scale puts on your feet and the downward push your feet put upon the scale. They're the same, thanks to Newton's third law. However, if you stand on the scale for one complete rotation of the Earth (a year), you will see the number change with the varying changes in force due to the Earth's movement.

Therefore, your terminal velocity depends on weight.

But hey, you want to hear something a little crazy? If you were to take that same scale with you into an elevator (which would be weird, granted, but just hang in there) and stand on it, you'll see the numbers change as the elevator accelerates and picks up speed. In fact, your weight will supposedly increase. Then, when you descend again and come to a stop, it'll read the same.

And then, if you happen to look at that scale while the elevator is still ascending but coming slowly to a stop, your weight will appear to be less. At the moment at which you begin a descent and the elevator speeds up but has not reached its peak speed, the same thing will happen again.

The biologist J. B. S. Haldane wrote about terminal velocity and gravity in a rather more jarring way. He stated that an animal the size of a mouse isn't going to retain any damage when falling to the bottom of a mine shaft. It'll get a small shock, then keep moving. Larger mammals such as a rat, a man, or a horse, are not going to survive a fall like that, however. This is because the amount of air resistance faced by a moving object is precisely comparative to the object's surface area.

The pushback against falling for the smaller mammal is 10x bigger than the force sending them downwards in the first place.

So, the bigger the object and the greater the fall, but keeping the same amount of resistance, the harder the impact. Ouch.

And on that note, on to friction!
If you try to move one object across another, and the two objects aren't made of the same material or have the same texture, you get **friction**. For example, when you try to push a heavy couch across a carpeted floor, it's really hard to do. That's friction! And when you hit the brakes on your car to stop at a red light, the reason you don't go careening through traffic anyway is due to friction!

You can climb rock walls, and you can start fires. You can wear those cozy socks with rubber on the bottom and not fall on your butt on a wood floor. All of this is because of friction. And you know what? In all of those instances, you are the heavier opposing force which makes the friction possible. No, I'm not calling you fat; it's true! The friction between two objects increases the heavier the opposing force, which is why those socks work: you're pressing your weight, sticky rubber and all, against the smooth floor. Likewise, your sneakers and groovy fingerless gloves gripping the rough wall and all of the foot and handholds are able to do so because of your opposing force, pulling you downwards.

Pertaining to mathematics, the force of friction looks like this:

$$F_{friction}$$

So if you're pushing or pulling something heavy, and you want to find out how quickly it can be moved in a certain direction, you take the dragging force without the power related to friction and equal that to the influence shifting in that direction. That looks like this:

$$F_{pull} - F_{friction} = ma \text{ (rate of acceleration)}$$

To calculate the amount of friction put upon an object, you first have to calculate the normal force, the force pressing down upon the object from above, preventing its movement. This is slightly different than the object's weight or the force due to gravity.

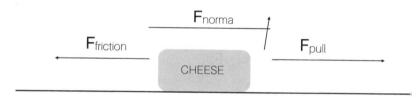

Okay, so imagine you're a mouse. You luck out one day and discover a block of cheese someone's dropped on the kitchen floor. If you happen to be a mouse who understands the concept of friction, you know it acts against the force you are going to use to try and push the cheese across the floor.

Lucky for you, in the case of just pushing something across the floor, the normal force (N) is going to be the same as the heaviness of the food, or:

$$F_{normal} = \text{weight}$$

From here, you find the quantity of power of friction. This will depend on what kind of surface your object is sitting on.

Luckily, physics equations tend to generalize as far as the type of surface and don't require different equations for carpet versus tile or anything like that. For example, to find the force of friction, you take the normal force and compare it to the friction force. In most cases, the two are similar, or even the same, so use symbol μ to represent the constant:

$$F_{friction} = \mu F_{normal}$$

The equation for actually finding the friction force, when you already have the normal force value, looks like this:

$$F_f = \mu F_n$$

That's the force of friction, which equals the constant times the normal force. So you multiply the normal force by its constant (μ) to get the amount of force. This is also called the **coefficient of friction**; this number will always be different depending on the type of surface you're working with.

The value of the coefficient usually comes between zero and one. A coefficient of zero is only achievable with absolutely no friction present at all, and such an environment is thought only to exist within a spatial vacuum, for which there is no proof of existence anywhere in our universe. If you wanted a coefficient greater than one, you'd have to get familiar with the world of drag racing (the kind with fast cars, not high heels). You have to build up quite a bit of friction to get up to speed in a shorter amount of time.

If you want to get even more specific, you can introduce **static friction** and **kinetic friction**. These two separate coefficients of friction take into consideration the different types of surface upon which friction is enacted.

When two objects are not moving but are pressed together, they can start to interlock on a microscopic level. They're static. When you slide one object on top of or across another, the two surfaces can't actually connect to one another because they're in motion. They're kinetic or mobile.

Static friction is the stronger of the two. This is obvious if you think about it because, with static friction, the two objects are at rest and have time to merge at the molecular level. On the other hand, kinetic friction ties two things together only at the macroscopical (usually visible to the naked eye) level because they don't have time to connect more deeply with one another.

So going back to the mouse and his giant block of cheese, say the static coefficient between the floor and the cheese is 0.3, and the cheese has a mass of about two pounds. And remember from earlier in this chapter, the formula for finding the coefficient of friction is Ff = μsFn. So you know the mass, and you know your surface (the floor) is flat. This also means the normal push (N) works in the opposing path than where the mouse is trying to go with its prize, and the value of the normal force is the same as the cheese's mass.

That formula looks like this: $F_f = \mu_s F n = \mu_s mg$.

Translation: The power of friction totals the fixed quantity of friction multiplied by the push, equaling the fixed quantity of friction times mass and acceleration.

The 'm' represents the cheese's mass, and 'g' is the increase of velocity because of gravity on the Earth's exterior.

The value **9.8 m/s²** is the average increase in speed of a dropping item due to the force of gravity on Earth, so by inputting the known values from our scenario, times that value, we come to the force of friction required by our little intrepid thief.

Go, little mouse, go! So now that he's got the proverbial ball rolling, how much force does he need to keep that cheese moving back towards his hole?

Well, that is where kinetic friction comes in. In order to get something to move, you must take it out of its state of static friction (rest), for example, by sliding a block of cheese across the floor. Once it is moving, the amount of force required to keep it moving is kinetic friction at work.

So our mouse needs 3.53 N to keep the cheese moving across the floor— hopefully, he had his Wheaties that morning!

In sports, the term "momentum" is frequently used. When an announcer says that a team has gained momentum, he or she is referring to the fact that it will take a lot of force to bring all of them to a complete standstill. The theory of momentum comes from general physics. Any object which shows to have any momentum will be difficult to bring to a halt. It is essential to employ a force counter to the movement of such an object for a specific period of time in order to bring it to a halt. The greater the amount of momentum that an object possesses, the more difficult it is to bring it to a halt. As a result, it would take more power or else a bigger measure of time or both to bring whatever this thing is to a complete stop. The object's velocity changes because of the power working on it for a specified period of time, and as a result, the entity's impetus changes as well.

If you've watched football in the past, you've almost certainly seen this before. In order to stymie the ball-carrier, the opposing players apply force for a specified duration. You've also had this happen while you were driving many times, right? For instance, the brakes are engaged in order to halt the vehicle in front of a stop sign or traffic light, applying a certain amount of force to the car to alter its forward momentum. When a drive is utilized to an item that is moving, the item will eventually come to a stop. If a push acts on an item for a set interval of time, the object's momentum will change.

When it comes to an unbalanced force, things speed up, no matter how fast they're going. An object's movement will be slowed when a force in the opposite direction to its movement is applied. In the event that a force compels an object to move in the same direction, that object will experience an increase in velocity. Thus, regardless of how it manifests itself, the force will alter the rate of that item. The item's momentum also changes if the object's velocity is altered. Newton's second principle was examined in a previous chapter, and these notions are an extension of it. In Newton's second principle ($F_{net} = m \cdot a$), objects accelerate based on the force applied to them and inversely on their mass (i.e., the acceleration is completely relational to the remaining energy working upon the item). This means that the equality of acceleration is obtained by combining its definition (a = change in velocity/time).

This equation signifies one of the two standards we will use in the evaluation of impacts, and it is expressed as follows. In order to fully appreciate the equation, it is necessary to comprehend its implication in words. It can also be stated that the amount of mass multiplied by the rate change equals the push times the passage of time. It is recognized as the impulse in physics and the "quantity force $*$ time" in mechanics. The quantity $m \cdot v$ must also represent the difference in momentum because m·v is a measure of momentum. The impulse equates to the shift in momentum, which is clear in the equation.

I hope to help you gain an understanding of the physics of collisions. When two objects collide, the laws of momentum govern their behavior, and the first law of momentum, which governs the physics of collisions, is expressed in the equation above. The impulse-momentum (I-M) change equation is the name given to this equation. The law can be stated in the following way: an object come into contact with an influence for a set quantity of time when it collides with another object, resulting in a transformation in momentum. When the force acts for an understood unit of time, the item's change in momentum is equal to the impulse that the object has experienced in the past.

Thus, **f • = m • v** is written in equation form.

When two objects collide, they both go through an instinct; the signal causes and equals the amount of change in the force of the objects. For example, consider the situation of a football player who runs down the football field when he comes into contact with a defensive back. Any confrontation with the opponent would alter the player's pace and, as a result, his forward movement. For example, if a ticker tape diagram represented the motion, it might look something like this:

. . . .
       ↑

The impact takes place approximately in the middle of the diagram and lasts for nine dots. When a halfback and defensive back crash into each other, the halfback will feel a force and will be set off balance for a period of time. The halfback slows down as a result of the collision, implying that the applied force came from the left. This would be the equivalent of a 600 N·s impulse if the halfback were struck by a 1200 N force for 1.2 seconds. The impulse caused this change in momentum, which is 720 kg·m/s. When two objects bump into each other, the impact on the momentum of each object is always equal. As an example, imagine the effect of a tennis ball hitting a wall. Ball rebound speed depends on both the physical characteristics of the ball and the wall. Likewise, the physical properties of the ball and the wall in question will vary when they collide. Every collision described above includes a bouncing ball colliding with a wall.

A larger rebound means a greater impact, and a greater change in momentum will lead to a greater impulse. After an impact, a collision can take place where the velocity and direction are both affected. Since the turn was made, the velocity has risen significantly. Objects that have the same or nearly the same speed prior to a rebound collision are likely to maintain that speed following the impact. Impacts in which things bounce back with the same velocity (and thus push and kinetic energy) after coming into contact with each other are known as elastic collisions. These are two relevant points about the mathematical disposition of the I-M change formula.

The first step is to remember that the impulse is the same as the difference in momentum. The statement is right, as the impulse is equal to force plus time. Another advantage is that since we already have two of these three quantities, we can determine the third one with relative ease. Furthermore, to conclude, since two of the last three rows are known, we can determine the row still in question. This is because the velocity of an entity is equivalent to the mass change multiplied by the change in momentum. There will be some further consideration about the subjective nature of the I-M Change Theorem.

For a given change in mass and speed, the impact force decreases by a factor of ten when the time of impact is increased. For the same duration and velocity, five times the mass is equal to five times the force. The effect of a double reduction in mass is compared to double growth in the rate of change with a constant force. A collision results in an object's mass changing direction as a force acts on it for a certain amount of time (i.e., where the outcome is a significant change in momentum). Before, we mentioned four material quantities: force, time, mass, and change in velocity. An impulse is the same as the force of an object multiplied by the time it takes to apply that force, and a change in momentum is the same as the mass of an object multiplied by its rate of speed. The ideas of momentum and sudden impulse are distinct. Objects never experience an impulse unless their momentum changes. The following equations convey this idea: This phenomenon is often referred to as the I-M change theorem. Let's now look at some examples of the I-M change theorem in the real world. That will be followed by some examples of physics at play in our everyday lives.

A good deal of what we will cover will be the effects of collision time and rebounding on the amount of force an object is put under. As you attempt to employ the I-M change theorem to a number of practical scenarios, look out for the following: The final aim is to make use of the equation as an example to working out how the amount of one variable quantity could be affected by changes in another variable.

Next, we'll examine how the magnitude of an object's collision force is impacted by the duration that it contacts the other object. I previously mentioned that the force-time relationship is inverse. You need to absorb 100 units of the momentum to stop a 100-unit-momentum object completely. If a user wishes to generate the 100 units of impulse required to stop an object moving at 100 units of momentum, he or she may choose from any amalgamation of force and time.

Time and Power Combinations Needed to Produce 100 Units of Momentum

Push	Moment	Stimulus
100	1	100
50	2	100
25	4	100
10	10	100
4	25	100
2	50	100
1	100	100
0.1	1000	100

The magnitude of the force acting on the object will be reduced as time passes between the collision and the object. It is important to increase the amount of time before impact occurs in order to reduce the force's effect on an object. It is imperative to shorten the time between collisions to get the best effect from an impact. A number of real-world applications for these phenomena are described below. Airbags in cars are a prime example. A car's airbags are employed because they extend the duration of the driver and passenger's safety in the car.

The first law of motion proposed by Newton states that people in a vehicle will be propelled forward after a collision. It's the change in momentum that launches them at the windshield, which then applies the needed force over a short period of time to stop their motion.

A driver and passenger might instead hit an airbag and extend the impact duration rather than the windshield. An object with some give, such as an airbag, could have its duration compounded by an aspect of 100 when struck. Multiplying the time by 100 makes the force go down by 100. That is a real-life illustration of physics.

The same idea can be used to explain why dashboards are cushioned. If the airbags don't go off (or if they haven't been properly installed inside of a vehicle), both the driver and passengers risk losing their momentum as a result of a clash with the windscreen or dash. Please keep in mind that the longer the period of time between the collision and the object, the less significant the force working on that object.
As a result, increasing the amount of time available to an object implicated in a crash is essential in order to reduce the influence of the force on the object involved. It should also be noted that, in order to augment the impact of a collision influence on an object participating in an accident, the time between collisions should be minimized.

A variety of real-life presentations for these occurrences are described in the following section. The use of airbags in automobiles is a good illustration of this concept. We use airbags in our cars because they decrease the amount of time it takes for the driver and passengers in the vehicle to come to a complete stop after an accident. In agreement with Newton's first law of motion, when a car collides with another, the driver and passenger have a tendency to continue moving forward as well. Their momentum propels them towards a windshield, resulting in a significant amount of force being applied all over a brief period of time in order to halt their forward movement and stop them from hitting the windshield. Instead of hitting the windshield, it is possible that the driver and passenger will collide with an airbag, which will decrease the length of time the impact lasts. When hitting an object with some give, such as an airbag, it is possible that the time duration will be multiplied by a factor of a hundred. As a result, multiplying the amount of time by a factor of 100 causes the amount of force you have to decrease by a factor of a hundred. That is a demonstration of the laws of physics in action.

In boxing, fans are often exposed to this same principle when fighters try and minimize the impact of a blow. Boxers will often loosen their neck and let their head move backward as they anticipate being hit in the head by their opponent. A boxer calls this "riding the punch." Boxers may try to prolong the contact of their glove with their head by leaning into the punch. The test is being lengthened to increase the time of the collision, which in turn decreases the force's impact on the collision.
If the collision time were raised by a factor of ten, the force on the object would be reduced by a factor of ten. The field of physics is working here.

Other sports, such as rock climbing, employ nylon ropes for the same reason they are used in basketball. For instance, rock climbers attach themselves to the cliffs' steep incline with nylon ropes. When a rock climber slips on a rock face, it will start to slide down it. The rope should stop her momentum in this scenario, thus preventing a fall to the ground that could cause her serious injury. Since nylon is so elastic, it is often used to make ropes. When a climber drops onto the rope, it stretches under the weight of the climber, causing a longer, more drawn-out force on the climber. To delay the moment the climber's momentum is lost makes it less likely for the climber to be injured when they fall. Climbers who want to minimize the effect of impact force can usually appreciate the merits of a longer delay between successive impacts—physics at work.

When hitting a ball with a racket or a bat, hitters are frequently encouraged to follow through with their strike. Researchers have discovered that following through with a ball while hitting it with a bat or a racket actually increases the amount of time it takes for a collision to occur. Matching to the I-M change theorem, any increase in time will usually influence one or more additional variables. Interestingly, the variable quantity which is reliant on time in this particular place is not the impending force. The force of a hit is determined by how hard and fast the batter turns the bat, and not because of however much time it takes for the ball to hit the ground. Instead, the follow-through lengthens the time between collisions and, as a result, contributes to a lengthening of the time between changes in the ball's speed.

The follow-through of the swing allows the hitter to bounce the ball off of the bat and increase the ball's speed. Of course, the faster the ball flies, or the harder the player hits, the greater likelihood of a successful play.

You're probably thinking of other examples of this principle right now. For example, you can throw and catch water balloons of different sizes and masses. If you throw a water balloon in the air, it lands on a target surface (i.e., grasped without rupturing). In order for the whole exercise to work, you have to catch the balloon, and therefore stop it from hitting the ground. If it then tumbles out of your hands and hits the ground, it's less likely to break since you interrupted its momentum. It's exactly the same strategy that lacrosse players use to catch the ball in their hands. When the ball is caught, it is "cradled," which means that the lacrosse player will grab for the ball and try to bring it inwards toward her frame, cradling it as she would a baby.. This lengthens the period of time during which the collision occurs, thereby reducing the influence applied to the lacrosse ball during the collision.

You can also observe this phenomenon by throwing an egg at a bedsheet. It is customary for two trustworthy students to hold the bed sheet while a volunteer throws the egg into it at breakneck speed from the edge of the bedsheet. Because the sheet is not totally solid, the collision between the egg and the sheet will last longer as the sheet bends to the force of the egg. If you were to hold the bedsheet completely taut, however, the egg would not be enveloped and thus slowed down. Instead, it would hit the sheet and bounce off, slowing a little but ultimately smacking onto the floor. The greater resistance, therefore, is an active component to the speed and drop of the egg.

Sometimes, two objects collide and do not either smash or slow to a stop. Sometimes, they will bounce into one another and pass one another, continuing in opposite directions at the same speed.

Based on the Theorem of Induced Momentum Change, we can deduce that the presence of an extremely large impulse is required in a rebounding situation. Therefore, a collision that concludes with a significant variation in momentum must also be associated with an equally substantial impulse because the impulse faced by an item is equal to the momentum change experienced by the object in the impact.

Rebounding is especially important in something like a car accident. When you get into an accident, two cars either collide and sort of bounce off one another and thus continue at a lesser velocity, or collide and crumple up and then come to a complete stop.

The thing is, in this case, the best-case scenario would be to immediately come to a complete stop. For one thing, if you go spinning off in another direction, you run the risk of getting hit or hitting someone else in another lane. If you hit all at once, however, odds are your car will absorb the impact.

Modern-day vehicles are actually designed better to take harder and harder hits and to protect those inside. Like airbags, specific areas of the car called crumple zones are made to give upon extreme force, such as another car or a deer. Believe it or not, allowing these sections to crumble instead of remain whole greatly increases your likelihood of survival.

When a part of the car— say, the front— crumples, it is gradually reducing your forward velocity. If it were to remain totally undamaged, both cars would stay at their previous speed, and the drivers would be propelled forward with much, much greater force. Instead, the car gradually slows, giving the airbags time to activate and slow your momentum even more. This prevents the trunk of the body from being impaled on the steering column or someone's head punching through the windshield.

For a less terrifying visual, take a soda can between your two hands, and see if you can crush it. Notice the way the can crumples, much like the car. If the can were not aluminum, but perhaps a glass bottle, if you tried hard enough, you could still break it, but it would shatter, rather than collapse, causing greater injury in a smaller amount of time.

CHAPTER 5
ELECTROMAGNETISM, WAVES, AND LIGHT

We'll start with the fundamentals here. To begin with: Waves! Waves are traveling disturbances that transfer energy but do not matter. Mechanical waves move away from where a medium or material's oscillations occur. This occurs because a rock placed in a body of water causes vertical waves, while the wave extends in all directions across the entire exterior of the water. Within the physics discipline, we will deal with two types of waves: transverse or diagonal waves and longitudinal or elongated waves. For example, a diagonal wave can be created by continuously jerking a rope side to side or upwards and downwards. The rope will alternate between rising (referred to as the crest) and falling (the trough) as it moves.

By principle, elongated waves ripple in a similar direction the further the wave moves. A long, soft spring, for instance, can be given a longitudinal wave by repeatedly pushing and pulling on it. These waves can be combined on occasion. One illustration of this might be waves in a pond, as the power of the wave's energy source is moving away from the power source, and the wave's reactions follow the wave. You may have previously heard about sound waves, which are vibrations in the air that go back and forth in longitudinal and perpendicular directions. The pressure variation, which our ears interpret, creates the sound that we hear as waves enter and pass through our inner ear. Wave characteristics such as amplitude, period, frequency, and wavelength can be described in several ways.

Amplitude, the first of these parameters, is the gap dividing the midpoint of a wave towards where the transposition is most significant. Consider the various machines you may have seen in hospitals that monitor heart rate. Waves on the screen move up and down to illustrate amplitude. All those wave patterns originate from the flat line, and the line's spikes correspond to the rate of the heart's beating. The displacement, as shown in this illustration, is measured as the spike's apex compared to the flatline. The amplitude of a wave is defined by the energy applied to it by its source. When you shake a rope, it affects the amplitude of the transverse wave it transmits. How much a loudspeaker or musical instrument compresses the air can determine the sound waves that are produced. There is no relationship between wave amplitude and its frequency, wavelength, or speed. The energy a wave holds varies on the amplitude and speed of the wave.

Waves can be categorized into two general groups: waves like those in water, sound, or EM waves, which spread over a wide area, and waves that stay in a narrow space, like fluctuations of a piece of rope or electrical pulses through a wire. The surface of a lake, pond, wide river, or ocean will often be covered with water waves. The spread of the energy source to other locations diminishes the power sent to a specific location as the source expands. Therefore, the water wave amplitude is similarly reduced as the distance from the source increases. In two dimensions, noise and EM impressions generally spread. Also, as the distance increases, amplitude decreases, but this time with a square relationship to the distance. This interval is the span between crest and trough and back to crest (see the hospital example). The cycle from flat to spike and back to flat again takes place in the span of one heartbeat. The wave's frequency is equal to the number of cycles it completes per second, and the wavelength is the distance from the wave's peak to its peak or trough to trough.

The frequency of a wave and the duration of a wave is always equal.

And if the length of time for any wave is a thousand seconds, this means the rate is 1/1000th cycles per second or 1/1000th of a Hertz (Hz). The wave regularity is equal to 1/period, and the period is equal to 1/wave frequency. It's the medium that governs the velocity of a wave. The more robust the interatomic/intermolecular bonds and the lighter the molecules in the medium, the quicker the wave will move.

The speed of all types of waves (transverse or longitudinal) moving through the same medium is the same. This is demonstrated by the fact that a sound wave in 0°C air travels at 1,085 feet per second, no matter the frequency or scale.

Wave velocity is dependent on both the water and wave frequency. Say you are constantly shaking a rope up and down. The speed of your hand in its motions is how many times it reaches the peak per second. The wavelength will remain constant as the waves travel along the rope. The wavelength is determined by both the oscillation frequency and the wave velocity. The relationship is given by $v = f\lambda$ or $\lambda = v/f$. It is important to remember that as the rate of a wave grows, the wavelength of that wave diminishes. Still, the speed of the wave does not change because the medium (the rope) has its own velocity.

The converse correlation between the rate and wavelength of any kind of wave can be stated as follows. While water waves appear to be transverse, they are a blend of slanting and elongated waves. Particles within any wave oscillate back and forth in increasingly smaller paths. Because the water wave's spherical path causes a moving surface, the wave has an undulating appearance. Of course, the wind is one key trigger of water waves. Because the water cannot match the speed of the wind, the water level is forced to rise and fall, giving the familiar wave-like motion. Waves are created in different sizes depending on the wind speed and how far the wind has been capable of traveling across the water.

Wave Type	Wind Speed	Results
Tube waves	< 3 knots	Small flows; the lengthier these ripples are made, the greater their scale
Rough waves	Three to twelve knots	Shared tube waves that have shifted and shaped even greater waves
Whitecaps	Eleven to fifteen knots	Scale of wave must be greater than 1/7 the wavelength for it to collapse into a whitecap.
Ocean billows	No measurable speed	Formed across great distances from crossed paths of different types of waves

An ocean wave's speed is based on the length of the distance between its crests or peaks. Longer wavelengths move faster than shorter wavelengths. Such as the ripples created by the wind, a small surface wave travels slowly due to its short wavelength. Greater waves caused by constant winds are longer and faster moving. Higher waves carry more energy than lower waves, and this explains why the former can inflict so much damage to the coastline. A cliff or shoreline that has mountains in the background generally does not get waves that break or develop whitecaps.

They break only when they reach a beachlike flatness, such as a shallow basin. A gradual rather than steep change in the shoreline depth will lead to a more dramatic whitecap. Waves break due to the effect of the wave velocity on the water's depth. Picture a large amplitude water wave. The wave begins moving toward the shore at a constant velocity. As the ocean gets shallower, the bottom of the wave begins to rub against the beach, which slows the bottom part of the wave relative to the top part. The crest passes over the trough while the bottom half of it slows down. When the crown can't be supported because of water scarcity, the wave either collapses or generates a whitecap. To continue, let's discuss a different type of wave: those pertaining to electromagnetism. EM waves are made up of two forms of waves. One oscillates within its particular field and the other moves perpendicular to the magnetic field. EM waves can travel through space without the help of any type of medium.

We'll cover EM waves as they pertain to light waves, as well as colors and hues, in the next chapter. EM waves exist in many different frequencies, referred to as the EM spectrum. The EM spectrum includes all types of EM radiation, including low-frequency radio waves to microwaves, infrared emission, visible light, UV emission, certain X-rays, and gamma rays. And as you may know, that last one is what supposedly triggered Bruce Banner's transformation into the Hulk, which is almost as impressive as your kitchen counter's little plastic box that can cook your potato in four minutes. By accelerating electrons, EM waves can be created.

Electrons generate a fluctuating EM field. The magnetic field is also oscillating, which creates an oscillating electric field, continuing this process. Waves emit energy that radiates into the surroundings, including the area around the moving charges. When it hits an electron-permissive material, it makes the particles oscillate (alternatively, "vary" or "fluctuate" if you want to remember it). We've used this word a lot recently.

James Clerk Maxwell, a Scottish mathematician, and physicist showed in 1861 how oscillating electric and magnetic fields are related mathematically. He introduced the actual concept of both magnetic and electrical areas using four specific calculations, later to be known as "Maxwell's Equations." According to Maxwell, it is due to the sequence of these four equations that the EM wave even exists. German physicist Heinrich Hertz actually proved Maxwell's theory of the existence of EM waves in 1886. He developed both a transmission apparatus and a type of receiver that could send and receive 4-meter waves. He measured wavelength using standing waves.

He demonstrated that their reflections, refractions, and polarizations could be produced and that they produced interference. In reality, it was the Hertz breakthroughs that helped establish the foundation for radio! Hertz was posthumously honored with the unit of frequency, the hertz (Hz), after his death in 1900. A wave that has no amplitude or frequency changes is incapable of carrying information, so it cannot be used for communication. One of the first ways to communicate using these waves was to toggle them in a repetitive pattern.

After an American inventor named Samuel F. B. Morse invented the code, which was used to transmit information over wires on a device known as the telegraph, characters were symbolized by long and short beats using what we now call Morse code. Flash forward to several decades later, when the telephone was invented by another American inventor, Alexander Graham Bell. His device used the same basic concept as the telegraph, but people could use their voices to communicate with one another over the wire instead of dots and dashes.

There's another guy, too! Italian inventor Guglielmo Marconi devised a radio device that could transmit and receive EM waves up to one kilometer away (3,280 feet). He received a British patent for his wireless telegraph after improvements to the device and the invention of an amplifier. He figured out how to send receiving set communications across the Atlantic Ocean in 1901 when he transmitted signals to ships 18 miles out to sea. Because of his contributions to radio transmitters and receivers, Marconi was later granted the Nobel Prize in Physics in 1909. And over the following decade, oceangoing ships installed and began using transmitters and receivers that had been significantly improved.

An antenna can be used to transfer and accept EM radio rays to and from radio and television signals. When the electrons in some kind of metal wire or rod—the transmittal projection—waver due to the vacillating voltages produced by a radio or television transmitter, they then create an equivocating electric field, which then, of course, is responsible for creating an equivocating magnetic field, and thus, another vacillating electric field. As the magnetic and electric waves leave the antenna, they move precisely at the constant velocity of light. A gathering antenna is usually some kind of metal rod, wire, or loop that can receive radio waves. When an EM wave touches the transmitter, the electrons in the metal vibrate in sync with the wave's frequency.

The receiver generates radio and television audio and visual signals due to the decoding of the oscillating electrons. EM waves are employed for communication in all frequency ranges in modern society. Numbers can range from kHz (representing a thousand hertz), MHz (representing a million hertz), and GHz (representing a billion hertz), depending on the size of the system and the distance between the sender and the receiver. If you remember, the name "hertz" is the number of oscillations or phases of the signal per second.

All transmitters create a certain frequency wave, which is called a carrier wave. When it comes to commercial broadcasters, the FCC designates the incidence. The transporter wave conveys no data.
To transmit data, the carrier wave has to have some of its properties altered or programmed in a way in order for a recipient can decipher or reconstruct the data. An analog encoding system uses a constant electric sign, while a numerical system splits the indication into numerous small-scale parts. Old radio broadcasting utilized analog signals coming from a microphone or other device.

Although commercial radio stations started out with analog technology, it was soon replaced by digital equipment. Both AM and FM are means of broadcasting information on the radio. The carrier wave's property is changed or modulated in each case. These various manifestations hold the data carried by the carrier wave. Since AM is easier to transmit and receive, it was the first method created.

In 1935, the American electrical engineer Edwin Howard Armstrong showed how FM could be transmitted and received. People were surprised by how FM eliminated "static" by completely changing the amplitude of a signal while preserving its frequency. Because the FM receiver's output is independent of signal amplitude, it is unaffected by interference. However, commercial FM broadcasts were delayed because the radio networks and receiver manufacturers were against it, and their implementation had to wait until after World War II. Today, music stations use FM channels for most of their broadcasts, while news programs and talk radio programs broadcast on AM for most of their programs.

Cell phones digitalize and transmit voice indicators. Specialized courses known as signaling processors condense any of the signals and add encoding to the transmission to correct transmission errors. Digital signals only transmit changes in order to perform compression. My guess is that you have heard all about the hype surrounding 2G, 4G, and 5G, as well as how cell phone companies keep making their cell service faster and more reliable while simultaneously allowing people to perform more tasks at once. But have you really figured out what that "G" means? All it is is "generation"!

The first generation of cell phones was only capable of connecting analog speech transmissions between two people. The evolving networks that came later allowed for a greater number of simultaneous users by employing digital methods and enabling email and text messaging on the device. As early as the year 2000, 3G networks made it possible for people to send and receive photos and videos. Also called the first real smartphone, 4G arrived soon after. This device, with its capability of GPS, map display, music, photo, and video sharing, was the advent of carrying something like a pocket-sized computer with you wherever you went. It's thanks to the gradual improvement of technology and the development of increasingly sophisticated network systems like CDMA, GSM, and finally LTE (long-term evolution). Mobile phones began as only able to make phone calls, and they gradually grew to be able to handle all of the apps, maps, videos, pictures, and games that we fill our smartphones with without a second thought. Both video conference calls (like Skype or Zoom) and live video recordings (for use on YouTube or Instagram) are included.

It's time to transition from our phones to food. It should be noted that, in part, this is also due to the fact that people are more likely to connect the microwave to the Internet through a wired rather than wireless connection.

The microwave EM waves with wavelengths between a few hundred microns and a few centimeters are more commonly known as microwaves. They are used as carrier waves for wireless communication at around 3 GHz. Microwaves of specific frequencies excite many common molecules and make them hot. The radiation that remains from the Big Bang, the microwave background, is all around us.

Microwaves are employed in the home for many different purposes. For example, laptops, wireless telephones, Bluetooth devices, satellite radio, and TV are all considered wireless internet devices. Another common use of microwaves is communicating between installations. Businesses, governments, and the military employ this technology for that purpose.

Microwaves can be broadcast in a few different ways, some more practical than others; line-of-sight transmission involves directing the microwave source at a microwave handset located no more than 18.64 miles away. The second method is to have the communication sent out towards a satellite which gathers it and transmits it back to a collecting plate. This is how you can have phone calls in the middle of nowhere, plus satellite radio and services like Netflix.

Microwaves are also produced in and transmitted by microwave ovens, of course. 2.4 GHz (2.4 billion hertz) waves are generated using a magnetron, and they are scattered throughout the oven. These microwaves vibrate and heat up water and fat molecules, causing them to spin faster and increasing their thermal energy.
The microwaves have different effects on various kinds of molecules, so some foods absorb the microwave energy differently (i.e., chicken broth is heated much faster than meatloaf; the liquid does not take as long to cook as the slab of solid beef).

For example, containers made specifically to be used in a microwave oven are constructed of substances that don't absorb the unsafe microwaves, so even when they help to cook the food, they don't warm up to the touch.

People who use microwave ovens would like to see the food cook while avoiding potentially dangerous microwaves. A grating with little openings is applied to deflect microwaves and replicate them back into the cooker to prevent their escape through the plastic or glass door. The microwaves, which are a little less than five inches long and are too large to get through the holes in the shield, can easily be seen by the visible light with its much smaller wavelength. Microwave radiation can potentially lead to hormonal imbalances, brain damage, malnutrition, weight gain, and a weakened immune system without that protective layer in place.

And while it's (ha!) great at protecting you, that grating should still be cleaned out regularly to make sure food particles and acidic liquids don't degrade it, which leaves you more susceptible to health issues. All that leftover mess from your reheated spaghetti sauce, TV dinners, or that middle-of-the-night brownie in a mug you made yourself in a low moment can actually break down that protective barrier specifically built into the microwave itself to protect you from the radiation it puts out.

CHAPTER 6
COLORS AND LIGHT

Do you know whether light travels at a finite or infinite speed? When we see the light, is it coming from the eye or going into it? In ancient Greece, Aristotle (who we've already discussed!) argued that light is not a moving object. According to Hero of Alexandria (10-70 C.E.), light actually does travel at an unlimited speed since you are able to see the stars as well as the Sun as soon as you tilt back your head and look up. Empedocles (490-430 B.C.E.) stated that it must move at a finite speed because it was in motion. According to Euclid (325-265 BCE) and Ptolemy (90-168 C.E.), in order for something to become visible, light is apparently sourced out of the eye itself. Many years later, in 1021 C.E., the Arabian researcher and philosopher Alhazen Ibn al-Haitham conducted experiments that supported the claim of a quantity of light moving from some kind of object and then flows inside the eye and so has to move at a fixed speed.

The Iranian polymath Abu Rayhan al-Biruni (973-1050) observed that light travels much faster than sound. Taqi al-Din (1521-1585), a Turkish astronomer, also argued for color theory and correctly explained reflection. Later, in the seventeenth century, the discipline of illumination made significant innovations. The German astrophysicist Johannes Kepler (1571-1630) and French logician, arithmetician, and physicist Rene Descartes (1596-1650) both supported one another's argument in favor of the constant speed of light, proposing that the sun, moon, and earth would not line up during a lunar eclipse if the rapidity of light was not endless. Despite this earlier mistake, Johannes Kepler would go on to define the theory of optics in his work, The Optical Study of Astronomy, in 1604.

He explained the inverse-square laws, how a pinhole camera works, and how surface and bowl-shaped mirrors reflect light. He was also able to identify the level of impact the atmosphere actually had on both eclipses and star positions. In 1621, Willebrord Snellius (1580-1626) (yes, that was his real name) ascertained the law of refraction (Snell's Law). Soon after, Descartes used Snell's Law to describe the development of rainbows. Christiaan Huygens (1629-1695), a Dutch scientist, authored several extremely valuable volumes on the science of optics and was one of the first who actually first voiced the idea of a specter of light actually being a type of wave.

Soon after, Isaac Newton's famous experiments with a prism where he would separate white light into its colors resulted in the publication of Newton's "Theory of Colors" in 1672. Newton accepted that telescope lenses produced tinted illustrations and devised the exhibiting telescope, which used a bowl-shaped mirror to avoid this flaw. Newton thought the light was composed of very light atoms known as corpuscles. In 1815 and 1818, the abstract and investigational work of French physicists Augustin-Jean Fresnel (1788-1827) and Simeon Poisson (1781-1840) solidified the current hypothesis of light. Later, James Clerk Maxwell (discussed in the previous chapter) and Heinrich Hertz (him as well!) demonstrated that detectable light is an EM wave. Starting from the initial few years of the 20th century, German physicists Max Planck (1858-1947) and Albert Einstein demonstrated that light, too, is a weightless unit. So, you could say that light and its waves have been the subject of discussion and debate for a very, very long time. On one hand, certain light possesses all of the assets of a crosswise wave, namely the ability to transport force and impetus. It follows the principle of superposition and can be diffused and inhibit itself. It does, however, have the properties of a massless particle. It starts in a running line at a continuous speed while in a medium. It has the ability to transfer energy and momentum. One of the basic ideas of quantum physics is a complete description of light's actual double life as both wave and particle (don't worry, we're not going to go there in this book). That's a little more advanced). The limits of the waves we call detectable light are defined by the limits of human vision. Light is commonly described in terms of wavelength rather than frequency because it was only possible to measure wavelengths and not light frequency until the last thirty years. The lowest visible light frequency is around four times the value of one thousand and fourteen hertz,, with a value of approximately 700 nanometers (700 x 10-9m). It appears to be a dark shade of scarlet to human eyes. The biggest frequency of light border is approximately 7.9×1014 Hz, corresponding to a length of four hundred nanometers. It appears to us to be a purple or violet color.

Infrared light has slightly longer wavelengths than red light, and UV light has wavelengths that are marginally shorter than violet light. You've probably noticed the light emanated by hot things, whether it's the overcast, burgundy glow of a range's heating coil, the orange glow of an oven's component, or the glow of an incandescent lamp's glowing filament. When light is then emitted by hot carbon particles causes the golden light of a fire. This type of illumination is commonly referred to as thermal radiation or black-body radiation. Thermal energy is created by converting energy, which is frequently derived from stored chemical energy. Radiation, including light, transfers that energy to the surrounding environment.

Unfortunately, this method of producing light is ineffective because approximately 97 percent of the released energy is infrared emission, which heats the atmosphere rather than ever producing any observable light. As a result of this discovery, numerous countries have started restricting incandescent lamps due to their high energy consumption. Fluorescence is the process by which vapors and objects emit light. Neon signs are an illustration of a gas that glows as a result of the conversion of electrical energy into light energy.

To produce intense light, some lamps use either sodium or mercury vapor. Electrical energy is used to excite mercury atoms in fluorescent tubes and compact fluorescent lights (C.F.L.s).

C.F.L.s are far more economical than fluorescent light bulbs, converting up to fifteen percent of whatever electric power they happen to consume into workable light. Today we use lasers in all kinds of unexpected ways, like in pointers in presentations, CD and DVD players, and even the handheld scanners at the grocery store. Usually they are made of small crystals, which in turn are a combination of gallium, arsenic, and aluminum. These laser beam out an intense ray of light, usually just in a single shade.

Lasers have the ability to shine a narrow light beam for an increased distance but can be dangerous if not used carefully. L.E.D. lights function using a kind of technology known as a light-emitting rectifying tube to produce brilliant light while emitting a lot less heat than traditional bulbs. They are built using the same materials as lasers, are capable of emitting light in different paths, rather than just a narrow beam. For decades, L.E.D.s have been used as control switch indicators in a number of appliances, traffic lights, vehicle lights, and television remote controls. L.E.D. light bulbs are now much more commonplace in households and enterprises alike; Although they are more expensive to manufacture, they are significantly more economical and last significantly longer than incandescent light bulbs. Because light carries energy, a light detector must convert the light energy into a different type of energy before it can be used to detect something.

Most of the time, light is transformed into raw electric energy.
For example, light strikes an opsin molecule in the human eye. Light absorption alters the structure of the particle, resulting in an electrical indicator being beamed along the optic nerve. When light is absorbed in a semiconductor in most computerized electrical sensors used today, such as the internal camera on your cellphone, any number of electrons get released. They generate a current as a result of the charge they carry, which is then translated into a numerical signal.

In 1638, Galileo Galilei proposed an approach to calculate the actual speed of light. In order to complete this experiment, Galileo would take a lamp, and there would be an assistant holding a second lamp that would be situated far away. Galileo's assistant was supposed to follow suit and expose his lamp once Galileo showed his. Following the administration of the flash to the assistant, the speed could be defined by quantifying the time it took for the light to move from Galileo to the associate and back again to Galileo's location. Galileo is said to have conducted the experiment in 1636, but no documentation has been discovered to support this claim. Two observers who were a mile apart participated in the experiment, which was conducted in 1667 by the academy of sciences in Florence, Italy.

Since there was no discernible interruption, they concluded that the speed of light must be exceptionally fast. Another scientist named Hippolyte Armond Fizeau (1819-1896) and was the first person to actually accurately measure the speed of light in an experimental setting in 1849. He makes use of a ray of light that passes through the spaces among the points of a fast-moving wheel, is displayed from a mirror eight kilometers away, and then bounces back to the wheel.

The wheel's rotational speed was enhanced until the restoring light could be seen passing through the next gap in the road. According to the calculations, the maximum speed was 315,000 kilometers per hour (19,5732 miles approximately). An improvement on this was made a year later by Leon Foucault (1819-1868), who decided he should use a revolving mirror instead of a kind of wheel and discovered the velocity to be 298,000 kilometers per second (18,5170 miles), which was less than one percent off the correct value at the time. Using this technique, he was also able to demonstrate that light moves more slowly in water than it does in air.

American physicist Albert Michelson significantly enhanced Foucault's dimension in 1879 simply by utilizing his own octagonal mirror on a rotating base and an additional, flat mirror located about 22 miles from a resource upon Mount Wilson in California. Michelson had the most contemporary quantity of the speed of light at that point in history.

The velocity of the mirror as it rotated and the separation of the two mirrors were both measured, producing an approximate speed of 299,910 kilometers per second (about 186,355 miles per second).

The theoretical speed of light could be calculated based on the interactions between power-driven rate and electrical area and between the electrical pulse and a magnetic area. Edward Bennett Rosa and Noah Ernest Dorsey arrived at 299,788 kilometers per second for the speed of light in 1907 using this approach. Michelson's measurements are as accurate as this one, both of which are within 0.1 percent.

Albert Michelson completed his new interferometric quantity of the velocity of light waves in 1926 that was accurate to 1/1000th of a percent. Many years after his initial work, exploration on microwaves used in the process of radar during WWII resulted in a whole new general application of his methods. Louis Essen and A. C. Gordon-

Smith described a calculation of 186,282 miles per second in the year 1950 using a hollow echo investigational method, and that turned out to be more detailed than Michelson's. K.D. Froome (1921-), when he used a radial interferometer in 1958 to try and measure the speed of light. Researchers at the National Institute of Standards and Technology in Boulder, Colorado, had successfully measured the wavelength and frequency of an infrared laser at the same time by the 1970s.

As the new standards were so exact, physicists altered the length standard definition. Krypton-86 emits a specific color of light, and the meter was previously defined as its wavelength length. The measurement of the speed of light was thought to be more precise than the definition of the term. As the new standards were so exact, physicists altered the length standard definition. Krypton-86 emits a specific color of light, and the meter was previously defined as its wavelength length. The measurement of the speed of light was thought to be more precise than the definition of the term. After that, the Conference Generale des Poids et Mesures (General Conference on Weights and Measures) made the decision in 1983 to set the understood standard for the velocity of light waves at 186,282 miles per second inside a controlled vacuum and defined the pattern as the space through which light will travel in 1/299792458 seconds. When waves travel through different materials, they travel at different speeds. On average, the light will typically travel at speeds of 299,792 km/s while in space. When the light confronts a fuller form, such as the Earth's environment, it will slow to 298,895 km/s. Whenever it strikes water, it decelerates considerably to 225,408 km/s or 3/4 of its original pace. Lastly, the light slows to 194,670 km/s when it passes through the dense medium of glass. The refractive index, denoted by the symbol n, is the comparison of the light's velocity within a medium compared to its unique speed inside a controlled vacuum.

Using lenses such as eyeglasses and microscopes, it is essential to pay attention to the refractive index. When any kind of light suddenly hits something, it can be immersed, transferred, or displayed. Things that are opaque absorb or reflect light. Light is both transmitted and reflected by materials that are either transparent or translucent. We should preserve the energy of light waves. The total energy absorbed, transmitted, and reflected by the material must equal the total energy striking it. The amount of light energy engrossed strengthens the thermal energy of the matter, which in turn causes it to heat up. The phenomenon of reflection occurs when light leaps off a surface, off of something like a looking glass or even a piece of newspaper. Mirrors and other smooth surfaces reflect light, corresponding to the established principle of reflection. This rule stipulates that the slant of frequency is equal to the angle of reflection. For example, if a light is directed at the surface at 30 degrees, it will be reflected at the same angle.

To put this theory to the test, you will need a mirror, a little flashlight, a piece of paper, and someone who is willing to assist you. Place the paper on the mirror and have your partner shine the flashlight on it. Once that's done, have them put the flashlight somewhere and use it to reflect light off your paper. You can try moving your paper closer to the flashlight's source, and changing the angle of the light's reflection, to see if you need to. After the gluing and drying, it's time to add some color! The phrase "white light" denotes a mixture of all the colors in the spectrum of visible light. Separating colors by wavelength produce different hues. Red has the longest wavelength, while the colors purple, blue, green, yellow, indigo, and orange have decreasing wavelengths. Out of all of these colors, purple (or violet) has the shortest wavelength. Most people cannot tell the difference between indigo and violet, and yet indigo is actually located somewhere between blue and violet on the color wheel. So, why are there seven colors when only six are present?

Historians of science contend that this correlation can be traced back to Isaac Newton, the scientist who penned the first treatise on optics and likened color to musical tones. The common European scale has seven notes: A, B, C, D, E, F, and G. As a result, Newton determined that the spectrum should have seven colors, with indigo being the seventh. But to most people, it can be extremely difficult to tell indigo from blue or violet. This is because some sort of light from an item has to actually enter ours in order for us to see it. We are able to observe stars, the strike of lightning, and light bulbs that have been turned on because they emit light. However, we depend on the light generated by these resources in order to see anything which doesn't produce its own light; we can only see these things because they reflect the light back to us, like the tree on the ground which gets lit up by the lightning.

This book's paper, for instance, does not have the ability to produce light. As opposed to reflecting light, the paper transmits it into our eyes, so we can view it. Because of this, when we think of colors, we are only picturing a small portion of the spectrum of white light. The visible light spectrum ends here. You have three color choices, as seen below:

The material allows you to view the object, although it blocks the remainder of the spectrum. The object's color itself. This means that it absorbs some spectrum and reflects others. It is also possible to physically separate the spectrum by diffraction or refraction, as in a prism or a rainbow.

Lighting "gels" (such as color filters) can be used on stage lights to produce various colored effects. While it is certainly true that green paper will reflect only the green portion of the spectrum, snow echoes the whole visible light spectrum, giving it a whitened appearance.
A black fabric will appear black since it absorbs all the visible light.

To show that colors were not created by the prism, Newton changed the experiment so that he could break apart white light into its component colors. If you were to go ahead and place some kind of lens in the center of the range, you could bring the colors together and concentrate them on another prism in the path of the colors.

The second prism definitely projected a shaft of white light. It would be impossible to distinguish between colors if the various wavelengths of light were all deflected by the identical sum as they entered and exited a prism.

All materials have their refractive index altered by light wavelength.
Red and blue have refractive indices of 1.571 and 1.594, respectively, in a diamond. The index in flint glass falls between 1.528 and 1.514. The crown glass ranges from 1.528 to 1.514, but only changes by 1.240 to 1.331 in water. In part, this is due to the fact that blue light has a higher refractive index than red light, causing it to be deflected through a greater angle than red light. The fact that diamonds sparkle is the reason for the noticeable differences in diamond quality. A state of complete darkness is diametrically opposed to white light, which is composed entirely of light or the absorption of all light.

This fact was discovered by Newton when it was entirely overlooked by everyone else. The paper is completely absorbing all of the light, so it appears to be black. Due to the fact that the sun is at a lower altitude in the sky and thus takes a more circuitous path through our atmosphere during the evening and early morning hours when it is low on the horizon, it takes longer for sunlight to reach us during those times of the day. Because more time passes for sunlight to travel a greater distance through the atmosphere in the daylight and dusk than it does during the day, more of the briefer wavelengths of light are dispersed away from the sun's direct light.

As a result, the sun's color changes from yellow to orange to red as the day progresses. The presence of dust and water vapor in the atmosphere causes sunsets to appear even redder, which is one of the reasons they are so beautiful. Another reason for the appearance of blue water in the ocean and other bodies of water is the presence of water molecules in the surrounding air. When looking at the water, the first thing that stands out is that it appears cloudy on cloudy days, whereas it appears transparent on sunny days. From one day to the next, a noticeable difference in the color of the water can be observed. Water will appear a deeper blue when the sky is clear and reflects it, as opposed to the water that appears a lighter blue when the sky is cloudy.

Additionally, water tends to scatter the blue wavelengths of light, which can cause bodies of water to appear blue. Orange, red, and long-wave infrared wavelengths are all absorbed by water. Because of this, it stores more solar energy, causing its temperature to rise—and more of the short-wave light is reflected, making the water appear bright blue. The spectrum of light known as a rainbow is formed when sunlight meets water droplets. When light enters a water droplet, it is refracted.

After the droplet's interior reflects light back against its rear surface, the light exits after refracting and dispersing. The angle between the entering and exiting light for blue light is 40 degrees, while it is 42 degrees for red light.

CHAPTER 7
PHYSICS IN THE REAL WORLD

So listen, we've covered all kinds of subjects and all levels of simple and convoluted concepts which apply to our natural world. Hopefully, I've made it abundantly clear that there is not a single corner of your life unaffected by the concepts of physics. Before we go, how about a refresher?

As a natural science, physics encompasses the study of matter, as well as the study of its movement and conduct through space and time.
It also includes the study of the bodies known as energy and force. Its primary goal is to gain a better sense of how the universe works, and its theories and sciences have been around for thousands of years to achieve this goal. It is possible that it is the oldest of all academic studies, particularly if astronomy is included in its scope. Most of the previous two millennia, physical science, chemistry, biology, and specific areas of mathematics have all been considered to be a part of natural philosophy or natural philosophy-inspired science. These natural sciences, on the other hand, were recognized as distinct investigation ventures in their own right throughout the Scientific Revolution, which occurred in the 17th century.

In many interdisciplinary research areas, such as biophysics and quantum chemistry, physics intersects with other fields of study, and the boundaries of physics are not delineated. Structural mechanisms explored by other sciences and new paths of research in scholastic disciplines such as mathematics and philosophy are frequently explained by new ideas in physics.

Advances in physics frequently pave the way for advancements in new technologies to be developed. For instance, developments in EM, solid-state physics, and nuclear physics have openly contributed to the expansion of new goods that have histrionically altered contemporary civilization, such as televisions and computers; advancements in thermodynamics have contributed to the growth of mechanization, and advancements in mechanics have contributed to the progress of the automobile industry. When it comes to describing the order found in the natural world, mathematics supports a succinct and precise language. All of the great thinkers of history, including Pythagoras, Plato, Galileo, and Newton, were aware of and supported this concept. To organize and formulate the results of experiments, physicists use mathematics to do so. Using experiments, researchers can find accurate and/or projected solutions, quantifiable outcomes from which new forecasts can be made and empirically verified or denied, and new predictions. Numerical data is generated as a result of physics experiments, and this data contains information such as unit conversions and estimates of measurement error. As a result of the advancement of mathematics-based technologies such as computation, computational physics is becoming increasingly popular.

For physics, ontology is required; however, for mathematics, ontology is not required. It follows that physics is ultimately involved with depictions of the real world, whereas mathematics is concerned with conceptual designs that go well outside the real world itself. As a result, statements in physics are synthetic, whereas statements in mathematics are analytic. Hypotheses are found in mathematics, whereas theories are found in physics. Mathematics statements must only be rationally accurate, whereas projections of physics reports must fit the studied and trial data in order to be valid.

The distinction is unambiguous, but it is not always apparent. Using mathematics in the field of physics, for example, is seriously just called mathematical physics. Although its techniques are mathematical, the matter is almost always physical.

For the most part, challenges in this field begin with a "mathematical model of a physical situation" (structure), followed by a "mathematical description of a physical law," which is then utilized within the system. All mathematical reports that are used to solve a problem have tangible connotations that are difficult to decipher. However, because it is exactly what the solver is looking for, the final mathematical solution has a more straightforward interpretation.

In the field of fundamental science (also known as basic science), pure physics is a subfield. Physics is referred to as "the fundamental science" because it governs all fields of natural science, including chemistry, astronomy, geology, and biology, and because the laws of physics apply to all of them.

In a similar vein, chemistry is frequently referred to as the "central science" for its part in linking the natural and life sciences. Consider the field of chemistry, which investigates matter's physical characteristics as well as its constructs and reactions (chemistry's emphasis on the molecular and atomic scale sets the field apart from physics). Networks are formed as a result of electrical forces between particles, properties are defined as the physical qualities of a given substance, and responses are governed by physical laws such as the preservation of energy, mass, and charge. Structures are formed as a result of electrical forces between particles, and properties are defined as the physical characteristics of a given substance. Physics is used in a variety of industries, including manufacturing and the practice of medicine.

Most scientists used the term "applied physics to refer to physics exploration that is meant for a specific function. A few classes in a related practical subject, such as geology or electrical engineering, are usually included in an affected physics degree program. When compared to engineers, an applied physicist is more likely to utilize physics or perform physics research to build new technologies or answer a problem rather than designing something specific in the traditional sense.

A similar approach to applied mathematics is taken in this case. Applied physicists are scientists who use physics in their research.
For example, in the case of accelerator physics, researchers may seek to improve particle detectors for use in theoretical physics research.
In engineering, physics is utilized extensively. Examples include the construction of bridges and other fixed structures, which require the use of statics, a subfield of mechanics. The knowledge and use of acoustics scores in improved sound control and concert halls; likewise, the application of optics results in improved optical devices.
A solid grasp of physics helps create more realistic flight simulators, video games, and movies and is frequently required in forensic investigations.

You can explore things that would typically be caught up in indecision because of the popular belief that the laws within the study of physics are static and are never changed or improved upon. Examples include studying Earth's formation, where it is possible to predict future or prior events by modeling the Earth's mass, temperature, and rotation rate as time functions. This allows one to generalize forward or backward in time and thus predict future or prior events. It also enables engineering simulations, which can significantly accelerate the development of new technological innovation.

However, because of the high degree of interdisciplinarity in physics, many other important fields (for example, the fields of econophysics and sociophysics) are influenced by the subject.

Physicists apply the scientific method to determine whether or not a physical theory is valid. Physicists are better skilled to assess the legitimacy of a hypothesis in a rational, impartial, and repeatable manner when they use a systematic methodology to evaluate the connotations of a theory with suppositions drawn from its trials and opinions. As a result, certain people conduct experiments, and go on to make their own observations in order to validate their theories.

Similarly, Newton's law of universal gravitation expresses a fundamental principle of some theory in a concise verbal or mathematical statement of a relationship, which is known as a scientific law.

Researchers strive to build mathematical patterns consistent with active trials, and that accurately foresee impending investigational findings, while experimenters design and carry out tests to test hypothetical projections and investigate new occurrences in the laboratory. Despite the fact that hypothesis and testing are established independently, they have a significant impact on and rely on one another. Experiments that defy existing theories, prompting intense focus on applicable modeling, and new theories that generate new experiments are two of the most common ways in which progress in physics happens (and often correlated gear).

Phenomenologists are physicists who work at the interface between theory and experiment. They study complicated inner workings of space and time, as well as other fields, and then attempt to apply them to specific theories.

The study of philosophy has stood to inspire the application and study of theoretical physics, and because of this, we discovered and learned how to manipulate electromagnetism.

Outside of the known universe, theoretical physics is concerned with theoretical questions, such as analogous universes, a multiverse, and superior dimensions. Theoreticians summon these ideas in the hopes of resolving specific problems with prevailing theories; they then investigate the outcomes of these theories and work toward the creation of testable estimates as a result of their exploration.

Engineering and technology both contribute to and are influenced by the expansion of experimental physics. For example, when conducting basic research, experimental physicists use paraphernalia such as particle accelerators and lasers. In contrast, those engaged in the practical investigation are more likely to be employed by trade, medical technology like magnetic resonance imaging (MRI), and transistors, among others. In addition, physicist Richard Feynman has observed that experimentalists may look for areas that theoretical researchers have not thoroughly explored.

Physics encompasses a wide variety of sensations, ranging from elementary particles (such as quarks, neutrinos, and electrons) to the most massive superclusters of galaxies in the universe. These experiences include the most fundamental objects that are the building blocks of all other things. As a result, physics is occasionally referred to as the "fundamental science." Furthermore, physics seeks to explain the many events that happen in nature by referring to humbler phenomena that occur in nature. As a result, physics aspires to connect the things that humans can observe to their underlying origins and then join these reasons together.

For example, in the primeval Chinese tradition, it was noted that an invisible force pulls together certain rocks (such as lodestone and magnetite). Magnetism is the term used to describe this phenomenon, which was initially meticulously investigated in the 17th century. While ancient Greeks were aware of other things such as amber, which, when scraped together, created a comparable unseen magnetism between the two, it was the Chinese who first discovered magnetism. This was also the first time thoroughly investigating it, and it was given the name electricity in the process. Physics had thus arrived at an understanding of two observations of nature as arising from a common source (electricity and magnetism).

However, when electromagnetism was discovered, it was discovered that these two powers were simply two different aspects of the same force—EMs. It is still going on now, and electromagnetics and the lesser nuclear forces are thought to be two facets of the electroweak interface rather than two separate entities.

The electroweak interaction, also known as the electroweak force, is a term applied in particle physics to explain the combination of two of the four recognized essential collaborations of nature: electromagnetism and the weak interface. Even though these two influences appear to be completely opposed at low energies encountered in day-to-day life, the theory shows them as two different phases of the same force.

In short, physics is attempting to discover the definitive reason (the theory of everything) for wherefore our environment is the way it is.

A broad classification of contemporary physics research can be made, with the most important subfields being nuclear and particle science; condensed matter science; atomic, molecular, and optical science; astrophysicists' research; and applied physics. Additionally, some physics divisions help physics education study as well as physics outreach.

From the 1900s, particular areas of physics have become progressively dedicated. Today, most contemporary physicists devote their entire careers to a single field. As a result, scientists known as "universalists," such as Albert Einstein (1879-1955) and Lev Landau (1908-1968), who performed in a variety of fields of physics, are becoming increasingly scarce.

Particle physics is the study of the fundamental components of matter and energy and the communications that occur amongst these constituents.

Also involved in this research are the designers and developers of high-energy accelerators, detectors, and computer programs that are required. Because many elementary particles do not occur spontaneously but rather are produced only when other particles collide with high energy, this field is also referred to as "high-energy physics."

Currently, the Standard Model is used to describe the exchanges of elementary particles and EM fields.

The standard considers the 12 common particles of matter (quarks and leptons) that co-operate with one another through the interactions of the powerful, vulnerable, and EM essential forces, among others.

The dynamics of matter particles are defined in terms of gauge bosons (gluons, W and Z bosons, and photons, to name a few) exchanging with one another.

In addition, the Standard Model forecasts the existence of a particle recognized as the Higgs boson.

Earlier this year, CERN, the European laboratory for particle physics, stated that it had discovered an atom coherent with the Higgs boson, which is a crucial component of the Higgs structure.

The field of nuclear physics is a specialized study of the interactions and components of atomic nuclei. It is one of the most well-known fields of physics. Nuclear physics is most associated with nuclear power production and nuclear weapons machinery. Still, the field has found applications in a wide range of fields, including nuclear medicine and a variety of other fields.

Atomic, molecular, and optical physics (AMO) has to do with the matter relations on the level of single atoms and molecules and interactions between light and matter at larger scales. They are often put in the same group because of how they are connected to one another, the common methods of dissecting them, and because they each possess similar energy to one another. Each of these areas includes classical, semi-classical, and quantum therapies in nature; they can treat their subjects from an atomic point of view (in disparity to a macroscopic view).

Atomic physics is the study of atoms' electron shells, which are made up of electrons. Even though the nucleus has an influence on atomic physics (see hyperfine splitting), intra-nuclear phenomena such as fission and fusion are believed to be part of nuclear physics.

The study of multi-atomic structures and their inner and exterior interactions with matter and light is the focus of molecular physics.

Optical physics focuses on the essential properties of visual fields and their communications with a matter in the microscopic space rather than on the control of classical light fields by macroscopic objects.

Condensed matter is a type of matter that has condensed into a solid-state.

This research is particularly interested in the "condensed" phases that occur when the number of atoms in a system is exceptionally large and their relations are extremely strong.

Solids and liquids are two of the most well-known examples of condensed phases, both of which are formed as a result of the bonding between atoms caused by EM force.

More reduced forms involve the viscosity of water and the Bose-Einstein condensate found in some atomic structures at extremely low temperatures, the superconducting phase demonstrated by conduction electrons in some objects ferromagnetic (permanent magnetism in metals) and antiferromagnetic.

Currently, the field of condensed matter physics is the greatest in the field of modern physics. Solid-state physics, which is now deemed to be one of the primary subfields of condensed matter physics, was historically the forerunner of condensed matter physics. Philip Anderson is credited with coining the term condensed matter physics in 1967 when he retitled his investigation group, which had formerly been known as solid-state theory. The American Physical Society (APS)'s Division of Solid Particle Physics was retitled the Division of

Condensed Matter Physics in 1978 after a long period of transition. Condensed matter physics has a great deal in common with other disciplines such as chemistry, materials science, nanotechnology, and engineering.

Astrophysics and astronomy are foremost the presentation of the theories and methods of physics to the study of stellar structure, stellar evolution, the beginning of the Solar System, and other questions of cosmology. Thus, astrophysics and astronomy are two distinct fields of study. Physicists who work in astrophysics typically draw on a wide range of disciplines, including mechanics, electromagnetism, statistical mechanics, thermodynamics, quantum mechanical theory and relativity, nuclear and particle physics, and organic chemistry, to complete their work.

Karl Jansky's discovery in 1931 that celestial bodies emitted radio signals sparked the development of a specialized science called radio astronomy. By being able to explore space more expansively, we are able to trace and map previously unknown structures within that realm. It is necessary to conduct infrared, UV, gamma-ray, and X-ray astronomy using space-based observations because the Earth's atmosphere causes perturbations and interference in the observations.

Physical cosmology is the analysis of the origin and progression of the world on the grandest possible ranges of time and space. The theory of relativity, which was ultimately established by Albert Einstein, has a crucial part in just about all current cosmological theories. For example, the discovery by Hubble in the early twentieth century that the universe is growing, as depicted by the map built by Hubble, jump-started the creation of many different theories, including that of the steady-state universe and the concept of the Big Bang. Nucleosynthesis within the cosmic microwave environment in 1964 provided conclusive evidence for this mysterious phenomenon.

The Big Bang model is supported by two theoretical pillars: general relativity, developed by Albert Einstein, and the astrophysical theory. In addition, astronomers have newly recognized the CDM paradigm of the development of our universe, which involves cosmic inflation, dark energy (which is not visible to the naked eye), and dark matter.

Fermi Gamma-ray Space Telescope data is expected to reveal a plethora of possibilities and discoveries over the next decade. These discoveries will fundamentally alter or elucidate current versions of the universe.

The possibility for a monumental breakthrough regarding dark matter, in particular, is exceptionally high over the next several centuries.

Fermi looks out for any feasible proof that dark matter is made up of enormous particles that interact weakly with one another, adding to similar experiments being conducted at the Large Hadron Collider and other belowground detection facilities.

The Meissner effect is quite possibly the best available example of a physical singularity that can be easily explained using common features of the study of physics: it is demonstrated by a magnet floating above a superconductor. One of the most compelling pieces of evidence supporting experimental physics comes in the form of indications that neutrinos have masses greater than zero. Because of these experimental results, it appears that the long-standing solar neutrino problem has been resolved, and the physics of massive neutrinos continues to be an extremely important subject of the latest and most dynamic academic and investigational examination.

The Higgs boson has already been tested at the Large Hadron Collider in Switzerland, where it has also, subsequently, already been confirmed. Efforts in the future will be directed toward proving or disproving supersymmetry, which, if you remember from an earlier chapter, is basically an add-on for the standard model for demystifying particle physics and is extremely favorable for the completion of contemporary experiments. Also currently underway are investigations into just what makes up the enigma called dark matter, as well as dark energy. These two things are considered to be of the utmost cosmic importance, yet scientists in the relevant fields have not yet been able to figure out what either of them is, or what they are made up, or how the heck they even came to be in the first place.

Despite significant progress in both quantum and astronomical physics in recent years, common but still misunderstood phenomena in our world remain unsolved, concerning certain complexities, elements of chaos, or indeed intergalactic turbulence. And it's frustrating to have still so much knowledge at your fingertips and to keep discovering more mysteries to interpret, within and without our known world. But it's exciting too, isn't it? To know, almost for certain, that we can never know the extent of all the wonders of our universe and the ones beyond it?

How do piles of sand form in the middle of a desolate desert? How do water nodes form? Those, by the way, are like silt deposits or collections of other water-based or non-water solvent materials found in shallow bodies of water, like creeks. Why are water droplets shaped like that? How, exactly, do the mechanics of surface tension, such as that of water on a leaf, or a duck's back, work?

Surface tension, by the way, is the lessening of the surface area of liquids to their smallest area possible when they are at rest. Creatures and objects such as razors and bugs, which are actually heavier than water (such insects as water bugs, which have narrow bodies and long legs in order to distribute their weight evenly), can float on water's surface because of surface tension.

All I'm trying to say is, physics is an endless field of study, involving every process and phenomenon on Earth and beyond.

I hope you have enjoyed this book and that it has provided you with a better understanding of how physics works in practice. No matter if you are a student, a teacher, or simply curious about the world around you, my goal with this volume was to open up the complicated and intimidating world of physics and to help you realize that it affects all of our lives on a daily basis and that even the most convoluted concepts have real-life applications and examples.

I hope you enjoy this book and find it useful. Thank you once again for taking the time to read this book, and I wish you the best of luck on your journey of discovery and future learning!

Hey! And don't miss this one! I'm sure that you will be interested:

Want to learn about the basics of quantum physics and impress your friends at cocktail parties with some "big brain" trivia about some of history's greatest scientific minds?

Think you could be a theoretical physicist, but you need to brush up on your knowledge of relativity first? Want to carry a book on the subway that will make your fellow passengers think you're totally sophisticated? Great! Then Quantum Physics for Beginners is the book for you.

Explore the field of quantum physics from its infancy through its bright future with topics like:

- Special and general relativity;
- The nature of classical physics v. quantum physics;
- What the heck is a quantum, anyway?
- Discovery of the atom and development of atomic models;
- Early experiments and research that changed the face of science forever;
- The photoelectric effect;
- Wave-particle duality;
- Schrödinger's contributions to physics (and his famous cats!);
- The life and works of Albert Einstein, including his 1905 'Miracle Year';
- The Heisenberg Uncertainty Principle;
- The Einstein-Bohr debates;
- Practical applications of quantum physics through the decades;
- Electromagnetic and gravitational waves;
- Unified field theory;
- And much more!

Quantum Physics for Beginners is a fun, sometimes irreverent look at the scientific concepts behind studying the behavior of the tiniest and largest things that we know to exist. With easy-to-understand analogies, this book looks at the big concepts, the groundbreaking theories, the brilliant scientists, and (sorry!) some of the complex mathematics that go into the study of how the universe works.

Read about how Einstein's theory of mass equivalency resulted in the world's most famous equation and how that equation eventually led to the creation of nuclear power and weapons. Learn about how Schrödinger never wanted to be the father of quantum mechanics but couldn't put the cat back into the box after his research was published.

Explore thought experiments and laboratory experiments that sparked debate and rapid scientific advancements throughout the 20th century. Quantum Physics for Beginners explains it all for you in layperson's terms so you can grasp the details of some of the wildest physics breakthroughs. It also explores the machinery necessary to measure and study the smallest atomic particles and most massive celestial bodies, all of which are governed by the laws of quantum physics.

Lastly, you'll take a journey through today's practical applications of quantum mechanics, chemistry, and physics to look at the future of clean energy, space travel, and medicine. You'll also be given a peek at the theoretical side of modern quantum physics and learn about the work that scientists are doing to make the impossible possible. Quantum Physics for Beginners will whet your appetite for studying how the world works and jog your brain into thinking about everything around you in a whole new way!

Made in United States
Troutdale, OR
12/18/2023